T0181584

We live in an era of complex and disjointed healthcare systems. Through the lens of the consumer, the practitioner, and work design strategist, this book provides innovative answers to vexed questions from various perspectives. The narratives are compelling, engaging, and informative.

Trajce Cvetkovski, *Discipline Leader in Occupational Health Safety and Environmental Management, Australian Catholic University*

The authors of *Healthcare Insights* share their unique perspectives in pursuit of a future that offers a more effective and compassionate healthcare system. From heart-breaking horrors to heroic success stories, the realities of healthcare unfold. The authors' lived experiences shed light on the system's complexity, its competing agendas, and often misunderstood or overlooked constraints that shape healthcare. Through a work design strategist lens, the stories also offer hope and inspires healthful decision-making and improved care practice. This book is a must-read for anyone keen to influence healthcare and seeking simple techniques that have powerful impacts on health and wellbeing. Aside from interest amongst work design strategists, health practitioners, health consumers, researchers, educators, and students this book is essential for business and government decision makers.

Elise Crawford, PhD (Human Factors), *Senior Lecturer of Occupational Health and Safety School of Health, Medical and Applied Sciences CQUniversity, Australia*

The Healthcare Insights book offers a thought-provoking connection between personal clinical journeys and the human factors and ergonomic (HFE) models and tools available to better understand 'work' and to improve clinical and system outcomes in these complex technical environments.

Sharon Todd, *Certified Professional Ergonomist and the current President of the HFESA*

Healthcare Insights

Uniquely, this book gives consumers a voice and regales tales of their experiences. These stories are complemented by the tales told by healthcare practitioners about their real-world constraints and evolving insights that have shifted their work focus. In the third section, work design strategists help the reader reimagine a better way to design the delivery of healthcare services and environments using human factors approaches. This interesting title:

- Covers real-world cases of people subject to an imperfect healthcare system
- Helps people understand the practical challenges affecting healthcare service delivery
- Champions new strategies to help people construct health, and to consider systems that will support these approaches
- Represents a broad array of healthcare settings

Healthcare Insights is well-suited to senior undergraduate, graduate students, practitioners, educators, and researchers in diverse fields, including healthcare administration and management, healthcare governance, human factors and ergonomics, service design, systems engineering, medicine, occupational health and return to work, allied health, work health and safety, workforce strategy, and architecture and design.

Sara Pazell is the principal work design strategist of a human factors consultancy practice. She is affiliated with five Australian universities for research and teaching in organisational strategy, work design, and occupational science.

Jo Boylan is Chief Executive at Clayton Church Homes. For the last 25 years, she has practiced and led authentically from a healthy settings approach. The settings approach drives health promotion and healthy ageing and involves maximizing early intervention and prevention to reduce risk.

Workplace Insights: Real-world Health, Safety, Wellbeing and Human Performance Cases

Series Editors: Nektarios Karanikas and Sara Pazell

The aim of the series is to host and disseminate real-world case studies at workplaces with a focus on balancing technical information. Further, the application of a work design framework will propel this series into the literary cross-over of traditional occupational health, safety, or wellbeing, human factors engineering, or organisational sciences, into a design realm like no other series has done. Each case will describe the tools and approaches applied per the Work (Re)Design stages and inform the readers with a complete picture and comprehensive understanding of the what's and why's of successful and "failed" attempts to improve the work health, safety, wellbeing, and performance within organizations.

Safety Insights
Success and Failure Stories of Practitioners
Edited by Nektarios Karanikas and Maria Mikela Chatzimichailidou

Ergonomic Insights
Successes and Failures of Work Design
Edited by Nektarios Karanikas and Sara Pazell

Healthcare Insights
The Voice of the Consumer, the Provider, and the Work Design Strategist
Edited by Sara Pazell and Jo Boylan

For more information on this series, please visit: https://www.routledge.com/Workplace-Insights/book-series/CRCWIRWHSWHPC

Healthcare Insights

The Voice of the Consumer, the Provider, and the Work Design Strategist

Edited by
Sara Pazell and Jo Boylan

Designed cover image: Katre Bailey, 45 Degrees Studio

First edition published 2024
by CRC Press
2385 NW Executive Center Drive, Suite 320, Boca Raton FL 33431

and by CRC Press
4 Park Square, Milton Park, Abingdon, Oxon, OX14 4RN

CRC Press is an imprint of Taylor & Francis Group, LLC

Library of Congress Cataloging-in-Publication Data
Names: Pazell, Sara, editor. | Boylan, Jo, editor.
Title: Healthcare insights : the voice of the consumer, the provider, and the work design strategist / edited by Sara Pazell and Jo Boylan.
Description: First edition. | Boca Raton FL : CRC Press, 2024. | Includes bibliographical references and index.
Identifiers: LCCN 2023043694 (print) | LCCN 2023043695 (ebook) | ISBN 9781032711867 (hardback) | ISBN 9781032422961 (paperback) | ISBN 9781032711195 (ebook)
Subjects: MESH: Delivery of Health Care | Sense of Coherence | Environment Design | Patient Satisfaction | Decision Making | Occupational Health | Personal Narrative
Classification: LCC RA425 (print) | LCC RA425 (ebook) | NLM W 84.1 | DDC 362.1—dc23/eng/20240102
LC record available at https://lccn.loc.gov/2023043694
LC ebook record available at https://lccn.loc.gov/2023043695

ISBN: 978-1-032-71186-7 (hbk)
ISBN: 978-1-032-42296-1 (pbk)
ISBN: 978-1-032-71119-5 (ebk)

DOI: 10.1201/9781032711195

Typeset in Times LT Std
by Apex CoVantage, LLC

'This book is dedicated to many people: my son, who inspired me to curate these stories; my mother, who always told me to dream big and turn my dreams into action; and to the many friends and family whose helping hands enabled me to stay buoyant during some of the most difficult moments and circumstances – this book is for you. I commend the contributing authors for their bravery and commitment to share their stories. The artful interplay of these compelling tales of struggle and dreams of a better world are buoyed by practical strategies to realise these ambitions'.

– Sara Pazell

Contents

PART II Practitioner Stories

Foreword

Healthcare plays an essential role in society, yet its practice is complex. Healthcare delivery is shaped by the perspective of the carers and care recipients, available resources, skill sets, and the environment of work. A fear of litigation and professional egos can affect practitioner decision-making. Within this landscape, consumers of healthcare may find themselves stumbling around as they seek the care that they need. Consumer empowerment is rare. In 2013, Queensland, Australia, established Ryan's Rule, a three-step escalation process that applies to patients admitted to all Queensland Health public hospitals, emergency departments, and, in some cases, 'Hospital in the Home' services (Clinical Excellence Queensland, 2023). Ryan's Rule was established via significant stakeholder collaboration. It was designed to empower patients, families, and carers to raise concerns and seek a second opinion if the patients' health condition worsened or failed to improve as expected. While it was a move in the right direction, better aligned care may be achieved early in the trajectory of care delivery if empirical evidence is sought. This includes listening carefully to the care recipient's human story, their life circumstances, and their accounts of symptoms. The impact of healthcare decisions extend well beyond the individual in need of care. Early collaboration and a greater understanding of stakeholders' needs and constraints can prompt more informed decision-making. This can lead to positive outcomes for those immediately impacted, and across entire social systems.

Unique to this book is that each chapter paints a picture of healthcare reality, and the authors openly share their stories in the hope that they will impact readers in ways that lead to better healthcare delivery. Towards this aim, the lived experiences of the authors provide perspectives of the consumer, the healthcare practitioner, and the work design strategist. Together, they give the reader a broader world view of the healthcare system. The chapters are organised to help the reader appreciate the realities of the care recipients, the challenges that they faced, and the opportunities available to healthcare providers and decision-makers about the organisation of work. There is hope because well-designed work and work environments can shape healthcare improvements. Together, these lived experiences enable the reader to adopt a broader perspective of the healthcare landscape and appreciate the matters that underpin healthcare decision-making. This vantage point lends itself to regenerative healthcare that moves beyond a focus on the disease state towards the construction of the health state, the effect of which benefits the key actors within the healthcare system and a vast social system.

The Workplace Insights book series (Karanikas & Pazell, 2021) describe real world experiences. The authors willingly and openly share their learnings whether gained through mindful effort or through their stumbling. This is a path that few will dare to tread, and it is what makes this book series remarkable. The authors of *Healthcare Insights* share their experiences: the good, the bad, and the ugly. These experiences are captivating, sometimes confronting, and difficult to read, while others are received easily and may have you in fits of laughter. You will be witness to the joy that was regained by aged care residents when activity directors learned the types of activities that they really preferred: beer at football matches, games nights, and dress-ups feature, noting that Bingo was rarely a preference.

Cases shared extend across all age groups and in all manner of situations. Many models of care are described and highlight difficult and sometimes shameful topics such as discrimination; ageism; mental health stigma; poor case management; and ignored calls for help, change, and reform. The authors acknowledge the realities of work and the need for carers to freely reflect, share, and promote learning without the fear of blame for matters often outside their control.

Unlike others, this book shows how healthcare is not all about problems or the diagnosis of disease but also about the creation and regeneration of health and well-being through a design lens. Simple techniques like authentic listening are found to have powerful effects, including the ability to circumvent the horrors of misdiagnosis while encouraging patient self-advocacy and personal resilience. The authors advocate for the role of Chief Work Design Strategist positioned at the executive level to influence reform across all aspects of care delivery.

This book, through science and empirical evidence, highlights the reasons why salutogenic approaches contribute to positive healthcare outcomes where healthcare is patient-centred and personal preferences matter. Remarkably, the editors of *Healthcare Insights*, Sara Pazell and Jo Boylan, masterfully thread together a wealth of real-world experiences to shed light on the realities of healthcare delivery. They show a mix of perspectives from many points of view which can create a ripple effect to mindfully construct healthful strategies. This book is a must read for healthcare and human factors practitioners, educators, researchers, and students. I recommend it for anyone seeking a brighter future in healthcare, whether for themselves, family, loved-ones, friends, neighbours, or workmates.

Dr Elise Crawford

REFERENCES

Clinical Excellence Queensland. (2023). *Ryan's rule*. https://clinicalexcellence.qld.gov.au/priority-areas/safety-and-quality/ryans-rule

Karanikas, N., & Pazell, S. (Eds.). (2021). *Workplace insights: Real-world health, safety, well-being and human performance cases*. Taylor & Francis Book Series.

Foreword

Healthcare ergonomics is a critical issue. The discipline of Human Factors/Ergonomics (HF/E) provides systems concepts and methods to improve care processes and outcomes for patients, caregivers, and clinicians. The science of fitting workplace conditions and job demands to the healthcare worker is paramount in these times where employee injuries, high turnover rate, increased sick days, and short staffing are only some of the factors that contribute to the stressors in our healthcare delivery systems. Health systems in Australia are under significant pressure to deliver high quality care against a backdrop of change in many disparate and competing domains, including patient demographics and care delivery, advances in treatment and technology development, and resources management.

In recent years, healthcare ergonomics has gained international recognition to provide direction for how managers establish healthcare delivery systems to ensure safety for patients and for healthcare providers. It has been called the 'bridge between quality and safety'. At the heart of these complex systems are our interdisciplinary healthcare teams and, combined, these can be viewed as 'complex adaptive systems' whereby good design makes a significant difference in enabling the provision of safe and effective health care.

I welcome this unique book, 'Healthcare Insights', the third within the Workplace Insights series, to help us understand the perspectives of consumers, practitioners, and human factors specialists/work system designers. To this end, I commend the authors in their art of weaving the stories synergistically.

Forward by Dr Thy Bao Thuy Do

Preface

A FEW WORDS FROM SARA

This book offers a novel approach to storytelling because many different people involved in healthcare systems share their perspectives. Readers can hear from consumers, practitioners, and designers. The consumers were brave in disclosing their humility as healthcare recipients, some of whom experienced harrowing and traumatising events. It is difficult to experience these events and it can be challenging, though sometimes cathartic, to reflect on them. The stories are relatable which may be unfortunate because they are all too common. It is confronting to learn that the age-old nightmare of waking mid surgery is possible today. It is saddening to hear about biases that affect care, such as those that marginalise women, older persons, or culturally and linguistically diverse people. Time and resource constraints are significant in this industry, and decisions may be compromised as a result. Unless a system supports better decision-making, mistakes will continue to be made. We thank our consumers for sharing their authentic stories openly, because our consumer-authors were hopeful about improving the experiences for others since much can be learned from their events.

In our consumer stories, we learn about what was supposed to be a routine colonoscopy procedure that turned into complications followed by an avalanche of cover-up of technical and human error. This led to the trauma of the patient, his wife, and his family. We continue to hear about medical error in the story about 'Mary', referred for cervical cancer treatments, yet patient-centred care was absent in her experience. We learn from Caponecchia who describes his family's experience while trying to support his father's care when language and cultural needs were disregarded in some instances. He explains the significant and adverse psychosocial impacts on his family caused by his father's long hospital stays.

The stories continue in the realm of paediatrics when misdiagnoses prevail, and confirmation biases occur when a general practitioner deferred to a 'specialist' to confirm a narrative that the parent refuted: this practice in medicine led to their omission of a routine, but important, diagnostic test. The trajectory of advised care could have caused the child's death if the mother had listened. Instead, she took matters mostly in her own hands while also seeking more holistic healthcare providers in a partnership with the child to reconstruct his gut health.

A chapter is provided in the form of a letter to healthcare administrators and governing bodies that was written by 'Miranda,' the daughter of an elderly man receiving hospital care describing the ageism and biases that almost caused their unnecessary transfer of Miranda's father to residential aged care when an alternative and effective return to home with rehabilitation care was possible. An effective return to home with rehabilitation was possible.

Practitioners were invited to participate in the storytelling because it must be acknowledged that there are real-world constraints to their effective work and models of care. Practitioners, as workers, must be cared for else who will care for the carers? These considerations often seem disregarded because of an unrealistic aspiration to deliver service to a healthcare consumer *at any cost*. That can be determinantal if a short-term goal to care for others (especially when it is monetarily incentivised) at any cost (emotional, psychological, or physical) creates fatigue and burnout among care providers. This book does not wish to *blame* workers but, rather, understand the realities of work that must be improved in healthcare systems. Skills and staffing shortages in healthcare are known as a global industry problem: these issues will not easily resolve. Practice innovations must be sought, and cogent design provides such avenues since technologies can be better leveraged, environments can be better crafted, and job roles can be better organised. Practitioners need spaces and places in which thinking and actions (mind and body) are unfettered, reflection is made possible, and confidence is restored. They need to rightly repel fear of being blamed for something that involves factors that are often beyond their control or that subject them to the vulnerabilities of the human condition. Workers, like healthcare consumers, need to feel safe and protected, and they need their health promoted. They deserve a good balance of meaningful and joyful work to counter draining circumstances.

The practitioners tell of stories about their journey as providers, as they evolve their expansive and holistic thinking about the creation of health. Beaumont tells of his emerging concept of health and how he has improved his role as a medical provider by understanding the shared relationship of care with healthcare consumers. Cavezza regales her story of burnout that led to her career redirection: by learning from her health needs, she has integrated her research into practice in functional medicine nutrition. Jajoo tells about her evolution from clinical practice to ergonomics design, starting in hospital systems. Henwood advocates passionately for exercise as an elixir to better health, especially for older people who must not forget this effective medium at any age (nor should the residential facilities that provide their care). Driver reminds us that effective healthcare outside the workplace impacts on timely and nurturing return to work. He reminds us that the workplace relies on these health-by-design partnerships. Ezbarami and Hassani inform readers about their pathways in human factors engineering, medical equipment design, and technology that can enhance healthcare delivery. They share some cultural experiences from their lives in Iran and their new pathway of work in Australia.

The book's work designers offer many progressive ideas about how to improve the healthcare experience for providers, consumers, and their families. Pazell tells of her early career experiences in healthcare clinical service delivery and executive management. She expounds on her advocacy for the appointment of a Chief Work Design Strategist in organisations. This is a position that can curate design practice and cultivate design partnerships that can impact on all aspects of operations and care delivery. Pazell, Meredith, and Hamilton describe their ideas on 'sensational work', or the reliance on sensory theory to recognise sensory needs and preferences of people and to inform the design of work. The healthcare setting is viewed as one in which these strategies offer significant opportunities to mitigate anxiety and duress, and improve healing and care. Sujan and Combes explain their work to deliver service along an elective care pathway and the impact of using human factors in design to address critical skills and staffing shortages in healthcare. They captivate readers by explaining their experiences of healthcare service and design during the pandemic.

Chari and Miller offer a liberating approach to healthcare and design research partnerships. They discuss their model of effective service improvement illustrated by case studies that can instil hope among those who feel that the framework of healthcare is fragile. Austin complements these ideas by describing the impact that human factors can have on emergency room service design improvements.

Bancroft is a guest reviewer because of her extensive experience in work health, safety governance, and transformative management. She is excited by what she has read in the chapters by Pazell, Chari and Miller, and Austin, and she is keen on implementing these ideas in her evolving executive roles.

The book is punctuated and concluded by the insights of Majd and Golembiewski who lead readers to marvel at the influence of a salutogenic orientation to healthcare: one that constructs health rather than focusses on mitigating disease and illness. They present the framework of salutogenesis and explain that small design changes in the built environment can have a profound impact on people's health and well-being; hence, these considerations are extremely important in environments in which health and healing is most needed. The authors provide an account of their personal stories to explain what has influenced them to think in these ways and how they came across the ideas that are now shared with our readers.

Importantly, design is beholden to empathy, because this is necessary to understand authentic needs, problems, and aspirations. From this foundation, building blocks can be formed to create better healthcare experiences for all involved. In my practice of human factors, human-centred strategies, ergonomics, and work design, I liken the evolving needs of healthcare systems to waterfall project management that requires formal planning combined with agile project management that requires the analytics of dynamic activities and measures to detect and rapidly regulate the system on a continual basis. For example, an academic model of a study design (which is common in human factors) might involve a highly organised and strategic approach (like waterfall management) with problems identified and needs articulated (often with a hypothesis unless undertaking qualitative research). There are analyses, intervention, testing, evaluation, and reflection. An agile model, however, is dynamic (akin to principles in resilience engineering): there is real-time, or near real-time data analysis with an appreciation that swift changes might be needed. Once performance expectations are established, parameters must be set around this with indicators on what detracts from these expectations. There must be early and cognisable alerts that provide responders with adequate response time. There must be an understanding of factors that prop performance, so that these can be readily reinforced, and some awareness of all involved in shared decision-making and agreed actions that impact on or improve the system. In safety sciences, this may be referred to as human factors engineering and resilience engineering. Of course, our authors have added to this by describing interventions, like nutrition and environmental design, that provide the right conditions for a person to heal thyself or thrive, depending on the needs; this can only enhance healthcare system design.

I am certain that there is much to learn from these stories: they have inspired me in my work and in the choices that I make for well living. The diversity and range of stories form the patchwork of a quilt decorated by shadow (sorrow or fear), light (hope), and colour (novel innovations). The pattern is human: our disappointments or despairs and, in many instances, our hopes and trusts. Some of the stories are alarming and confronting, but what is inspiring to me is the realm of possibilities for healthcare design improvements. I believe, strongly, that we must consider the design of work from a systems' perspective of service design and user experience. The work of Neumann and Purdy (2023) describe a better care framework with seven strategies for sustainable healthcare system process improvement: setting integrated goals, using key performance indicators, active stakeholder engagement, using human-centred design thinking, fostering organisational learning, taking a stance of development, and respecting local priorities. They discuss a multi-factorial approach to impact on operations and clinical care outcomes, including staffing concerns, logistics, technology, the built environment, and system design (Neumann & Purdy, 2023). I believe the authors of this *Healthcare Insights* book have addressed many of these ideas in the checkerboard of their stories. The consumers, the practitioners, and the work design strategists advocated for system improvements to support care delivery and to support family and carers.

I thank our contributing authors and, importantly, I thank Jo Boylan who inspired me in my early healthcare education and career to embrace all life transitions, at any age, as ones that offer many moments of well living, if only we design for and seize the opportunities.

REFERENCE

Neumann, W. P., & Purdy, N. (2023). The better work, better care framework: 7 strategies for sustainable healthcare system process improvement. *Health Systems*, 1–17 (ahead-of-print). https://doi.org/10.1080/20476965.2023.2198580

A FEW WORDS FROM JO

Self-advocacy Builds Resilience

The chapters in this book have inspired me to write from the heart and to tell my story of challenge, advocacy, and resilience during my work in aged care. I was moved by the highly resilient authors who found the courage to change course when needed or to adapt to difficult situations (often, despite the advice of a 'voice of authority'). They built their emotional resilience and reinvented themselves to achieve their health aspirations.

In many of these stories, resilience arose from adversity and the extraordinary self-advocacy that was required to speak up about what mattered. Self-advocacy can build resilience and I believe that it constructs a mindset of learning and continuous improvement. Thankfully, self-advocacy and resilience are teachable and arguably necessary for people working in or interacting with the healthcare system. Without being informed, decisive, flexible, supportive, and confident to lead, these stories may have ended differently. Barriers to process improvements and outcome-focused care were evident in many of the stories. However, flexibility, adaptability, and perseverance prevailed to achieve better outcomes.

To ensure that processes, compliance-based procedures, and controls are in place, there must be considerable governance oversight which can be faulty or remiss if monitoring is not systematised. Dr W. Deming's (1991) suggested 'a bad system will beat a good person every time'. He highlighted the importance of blaming the process and not the person when things go wrong to avoid losing employee engagement in future process improvement.

Neumann and Purdy's (2023) 'Better Work Better Care Framework' is a great example of how healthcare can adopt a systems-based approach to sustainable care delivery. This framework outlines strategies to design, manage, operate, and achieve improvement in healthcare. The adoption of this type of framework is fundamental in healthcare to support managers and change agents in effective care delivery and to prevent adversity or harm.

I thank our contributing authors and, importantly, I thank Sara Pazell who has coordinated and motivated us all to share our knowledge through compelling stories that grasp the narrative around many of today's key challenges. These chapters help us make sense of the world and serve to educate, inform, connect, inspire, and persuade others to join the advocacy efforts and lead improvements to achieve the best possible outcomes in healthcare delivery.

REFERENCES

Deming, W. E. (1991). *The Deming institute*. Retrieved June 23, 2023, from www.quotes.deming.org/10091

Neumann, W. P., & Purdy, N. (2023). The better work, better care framework: 7 strategies for sustainable healthcare system process improvement. *Health Systems*. https://doi.org/10.1080/20476965.2023.2198580

Editors

Sara Pazell

Dr Sara Pazell is the principal work design strategist of a human factors consultancy practice. She is affiliated with five Australian universities for research and teaching in organisational strategy, work design, and occupational science. She is a participatory, transformative, action researcher, relying on science and shared learning to solve real-world problems in partnership with those who are most impacted by design changes. Her interests are in organisational strategy, human-centred design, and human factors in systems design. She has held past roles as a healthcare executive and operations manager in varied settings. Sara manages a household of boys (one son, one dog, and a sassy cat).

Sara is a series co-editor for the Workplace Insights book series and book co-editor for the Ergonomic Insights book in the series. She advises on national and global committees on health, well-being, and design. She has been selected to serve on innovation, leadership, and design committees globally. Sara is a rabble-rouser on the globally recognised and popular WhyWork Podcast.

Josephine (Jo) Boylan

For the last 25 years, I have practised and led authentically from a healthy settings approach. The settings approach drives health promotion and healthy ageing and involves maximising early intervention and prevention to reduce the risk of fraily among care recipients.

I am a strong advocate for Australian Aged Care Services to recognise the restorative potential of older adults. During the past 20 years, I designed systems and services (structures, governance, and practices) to make healthy normal for even our oldest residents or clients.

My educational background is in nursing and public health, establishing my skills and confidence to drive good clinical governance and practices towards early intervention and prevention.

More recently, I am the Chief Executive at Clayton Church Homes and our Board is investing in a Positive Ageing vision, including establishing gyms and exercise physiology in each of our homes and services. This enables the residents access to the resources that will help them push back on avoidable decline and disability in their later years.

Contributors

Amy, Amy is a sole parent who has juggled many roles while honouring the most central obligations: care of her son and provision of a safe, secure, pet-warmed, and loving household.

Elizabeth Austin, BA-Psychology (Hons), PhD, is a human factors and resilience researcher at the Australian Institute of Health Innovation, Macquarie University Sydney. Dr Austin leads interdisciplinary research integrating psychological theories, human factors analysis, and systems theory to improve quality, safety, and sustainability in complex socio-technical systems, with a focus on Emergency Departments.

Kym Bancroft is a passionate, future state-driven health and safety professional with 20 years of experience strategically leading safety, health, well-being, environment, sustainability, and injury management functions across a diverse range of global organisations, achieving safe, fair, productive, and sustainable outcomes. Kym's experience includes senior roles as a government regulator and head of health, safety, and environment for global public services and state utilities organisations Serco Asia Pacific and Urban Utilities. Kym has a master's degree in applied psychology (organisational) and a master's degree in safety leadership.

David Beaumont is Consultant Occupational Physician, a doctor who specialises in the health of workers. His background was in general practice before specialising in occupational medicine. A former president of the Australasian Faculty of Occupational and Environmental Medicine of the Royal Australasian College of Physicians, he is the author of *Positive Medicine: Disrupting the Future of Medical Practice.*

Carlo Caponecchia is Associate Professor in the School of Aviation at the University of New South Wales (UNSW) Sydney and Co-associate Dean Equity Diversity and Inclusion in the Faculty of Science, UNSW. Carlo's specialities are in psychology, human factors, and safety, and he was part of the international standards technical working group for the development of the international standard on psychological health and safety at work (ISO45003).

Shelly Cavezza is a scientist with a PhD in BioMedical Science from the University of Sydney and over 20 years' experience as a Principal Research Scientist. Her specialities are in functional medicine and clinical practice in translational nutrigenomics, human nutrition, and personal DNA profiling. She led research and authored more than 80 peer-review papers in scientific journals on molecular and cell biology, immunology, allergy, genetics, functional nutrition, and natural therapies.

Satyan Chari is an accomplished occupational therapist, human factors researcher, and healthcare quality and safety leader. Satyan's doctorate studies were on patient safety and a human-centred environmental design to reduce falls and improve recovery among hospitalised, older patients. Satyan led the Clinical Excellence Queensland Bridge Labs award-winning innovation programme for tackling complex frontline healthcare challenges through government–academic partnerships in creative design and work systems. He is affiliated with several Australian universities for teaching and research.

Julie Combes, RSCN, BSc (Hons), MA Clin Ed, FHEA, is Deputy Director of Elect consults at NHS Elect, leading implementation of workforce and training innovations into clinical practice for improvement. Julie started working in the NHS back in 2000; she began her NHS career as a paediatric nurse, going on to become an intensive care nurse. She has been involved in workforce education and training in the NHS since 2010. With special interest and expertise in interprofessional education, patient safety, and human factors, she is an associate member of Chartered Institute of Ergonomics and Human Factors, working in partnership to integrate human factors science into healthcare.

Elise Crawford is Senior Lecturer within the Safety Sciences discipline at CQUniversity, Australia, who specialises in human factors. She is a work design advocate who promotes human-centred and participatory design to advance sustainable change through adaptability, innovation, health, safety, and well-being.

Thy Do is an Honorary Research Fellow, School of Psychological Sciences, University of Western Australia and the Director of Operations, Nexus Foundational Human Factors Training for Healthcare, Royal Perth Bentley Group, East Metropolitan Health Service, Perth, Western Australia. She is a member of the Australian and New Zealand College of Anaesthetists and the Chair, HealthSIG, of the Human Factors and Ergonomics Society of Australia (HFESA).

Jane and John Doe are a married couple who run a household of children propped by love and hard work. Their union was strengthened by their battles and woes in their conflicting roles as a healthcare provider and healthcare recipient.

Nicholas Driver is an accomplished OHS practitioner in Queensland (Australia), with almost 20 years' experience in Occupational Health and Safety, and Workers' Compensation. He has a bachelor's degree (with honours) in Commerce majoring in Organisational Behaviour and Information Systems Management, and a Diploma in Occupational Health and Safety. He has operated across multiple contexts, including state and local government; private enterprise; and primary, secondary, tertiary, and special education settings.

Sahebeh Mirzaei Ezbarami is a researcher in the field of medical engineering with over five years of professional experience. She has worked on numerous projects related to occupational health/safety and medical devices. Her interest concentrated on designing and implementing innovative solutions that enhance workplace safety and mitigate health risks.

Jan A. Golembiewski is a practising architect and Director of Psychological Design, a specialist architectural firm working out of Sydney, Australia. As an architect and educator, Golembiewski is recognised for his work specialising in health-related projects, human-centred innovation, design psychology, salutogenic health promotion, ecological psychology, and design for behavioural, affective, and psychological reactions to the physical environment. Jan is also involved in a startup company to bring Earthbuilt, a new carbon-free, inexpensive building material to market.

Anita Hamilton is Senior Lecturer and current Program Coordinator of the Occupational Therapy Discipline at the University of Sunshine Coast. Anita has an extensive clinical background as an occupational therapist in mental health, work rehabilitation, and acute hospital services. Anita's PhD explored digital literacy in the occupational therapy profession and since then she has developed three distinct research portfolios: (1) Animal-assisted Interventions, (2) Well-being, and (3) Scholarship of Learning and Teaching in Higher Education. Anita is a mother of two adult children and became a 'Nanita' to twin granddaughters in 2022. Anita and her husband share their life with their dog Cooper, who is also a therapy dog.

Mehrdad Hassani is a researcher in the field of occupational health, human factors, and artificial intelligence.

Tim Henwood is the Group Manager, Health & Wellness, Southern Cross Care (SA, NT & VIC) and Adjunct Research Fellow at the University of Queensland. He has a passion for healthy ageing and evidence-based models of care that put older adults onto better health pathways. His career in healthy ageing started as an academic where he spent 20 years and published over 85 peer-reviewed papers that demonstrated the value of physical activity for functional well-being in community and residential aged care adults.

Bharati Jajoo is a leading ergonomic consultant and the founder of Body Dynamics in Bangalore. India. She has pioneered, developed, and managed corporate in-house ergonomic/wellness programmes for India's leading corporate houses for many years. Her work experiences range from an acute hospital inpatient therapist to treating musculoskeletal disorders in the United States and India.

Narges Farahnak Majd holds a master's degree in architecture and she is a PhD student at the Queensland University of Technology. She is a researcher in architectural psychology. Her research interests include mental health and architecture, learning spaces design, the application of neuroscience in urban design and architecture, and healthcare and therapeutic design.

Mary is a sole parent, mother of two, children's author and researcher, affiliated with several Australian universities. Mary's other interests include yoga and long meandering walks on the beach.

Pamela Meredith is Professor and Discipline Lead of Occupational Therapy at the University of the Sunshine Coast and is Fellow of the Occupational Therapy Australia Research Academy (FOTARA). Pamela is an occupational therapist and a psychologist. She has a demonstrated track record in the publication and dissemination of her research activity with more than 120 publications. Pamela is also a dedicated Research Higher Degree Supervisor. She lives with her husband, two young men, and a menagerie.

Evonne Miller is Professor of Design Psychology and Director of the Design Lab at Queensland University of Technology. Her research centres on design for health, and she is Co-Director of the Queensland Health–Funded HEAL (Healthcare Excellence AcceLerator) initiative, which embeds designers, and 'design doing and thinking', into healthcare.

Miranda, Miranda is a working professional who has twice helped intervene in her father's care, once helping to prevent the wrong surgery from occurring (that could have caused her father severe spinal cord damage), and the other to improve his care coordination, as was told in this book.

Mark Sujan leads research and consultancy projects on human factors in safety-critical industries, such as healthcare, process industries, and air traffic management. He is Chartered Ergonomist and Human Factors Specialist (C. ErgHF) and Fellow of the Safety and Reliability Society (FSaRS). Mark works part-time in the NHS, providing education in safety investigation. He was Trustee of the Chartered Institute of Ergonomics & Human Factors (CIEHF), where he chairs the Digital Health & AI special interest group.

Part I

Consumer Stories

1 When Alarms No Longer Cause Alarm

Jane and John Doe

1.1 PATIENT STORY

The following incident took place during what was supposed to be a routine colonoscopy procedure. I had undergone these procedures biennially for 20 years: this would have been my tenth 'routine procedure'. Many people experience this medical procedure and in Australia more than 800,000 were performed in 2016–2017, or 1 in every 32 Australians (Swannel, 2018). My wife dropped me off in the morning. It started as a calm day before my surgery. I was looking forward to the deep sleep from anaesthesia. My wife dropped me off in the morning. She had planned to collect me a few hours later, and I expected that we would go about our typical day. However, this is not how the day unfolded.

Shortly after recovering from the colonoscopy, the doctor came to see me with news of a potential perforation to the bowel that occurred during the procedure. This meant that I needed to stay in hospital overnight for monitoring. He explained that the worst-case scenario would be another procedure to repair the damage. That evening, as the painkillers began to wear off, I felt the worst abdominal pain that I had ever experienced. It was to the point that I almost blacked out. I learned later that this was a sign of the worst-case scenario, and the care team took me straight to the operating theatre. I recall being in the operating theatre and going under general anaesthesia again.

During the procedure, I remember waking up. I initially thought that I was in the recovery room. Waking from the surgery, I heard people talking and I saw the torsos of the doctors and nurses around my lower body. I felt pulling and tugging in my abdominal area and I felt pain. At this stage, I realised that they were still operating on me. I began calling out, 'I am awake!' and 'I can feel that – stop – stop!' I started to scream, thinking that they could not hear me. I begged them to stop. I realised that they could not hear me. For a moment, I felt like I was hovering over my body and I thought, 'This is it, I am dying'. I realised that I was paralysed. I was becoming more distressed and anxious. Could they not see that my eyes were open? I thought, 'I can see you. Can you see me?' I waited for the angels to appear. Where were they? I did not get to say to my loved ones.

I recall struggling and then waking up in the recovery room. My wife was sitting next to me. I thought, 'Maybe I am alive'.

1.2 POST-INCIDENT

The physical recovery post-surgery was tough, coupled with fatigue and nightmares for the first five days. I found it hard to sleep. *How can things go so terribly wrong? Why*

DOI: 10.1201/9781032711195-2

me? Out of all people, why ME? I had had a complication from a routine colonoscopy. It escalated into emergency procedures, from which I woke mid-abdominal surgery. I was traumatised by this, and it took some time before anyone realised that this had happened.

They put me in the best room in the ward. The staff checked on me regularly. Yes they treated me well afterwards. I was a private patient. I wondered if, in the public system, the staff would have afforded me the same post-care. However, all I cared about was that this error did not recur. I could see it happening in both public and private sectors. It made me question why I was paying private fees. I paid for quality care to avoid these situations. Serious errors can happen regardless. Money can only do so much. In the initial aftermath of the incident, my wife requested a meeting with the hospital director. The hospital was initially responsive and there was an effort to collaborate on procedural strategies to prevent another patient's experience of waking mid-surgery. However, all hospital management contact ceased six weeks later. This wasn't reassuring because I would have like to have seen the improvement strategies and, thus, have confidence in the system again. I don't feel ill about the doctors involved and continue to consult my specialist. However, I just hope that I do not need to fight this, advocate for myself, convert this to a legal battle, reflect on this for much longer, deal with the emotional fallout of trauma, or put my family through the management of any other related complications.

1.3 HOSPITAL DISCHARGE AND GETTING BACK TO LIFE

The following weeks after hospital discharge were tough, and I experienced several recurring flashbacks. One is the recall of my surgery, the others involve walking through dark corridors in hospitals, seeing sick and bandaged patients. The feeling was eerie. For a while, I could not watch any medical shows on TV. Beeping noises and alarms would cause me anxiety. My sleep and concentration became difficult, which lingered for months, and my work performance suffered noticeably. I resigned from work.

My wife called providers for two weeks straight, even using her personal network to book me into see a psychologist. But no success. Waiting lists were long and it would be six weeks before I could start seeing a psychologist. Why was it so hard to find someone?

1.4 CONTROL

Months later, I started to argue with my wife. 'Stop trying to control me', she would say. 'You tell me where to stand in line at the supermarkets, and you tell me when we need to leave. You tell me which coffee shop to buy the coffee, it is always at your direction'. I thought that what I was doing was normal: most guys hate shopping. I still having issues letting go of control when I need to. It's going to rain, so I won't go out today . . . so . . . it seems that every cloud doesn't have a silver lining.

1.5 LOSING EMPATHY

For the first 12 months, I kept a distance from any family or friends especially when they had concerning issues to discuss. I could not deal with any more drama, so I slowly cut off people. I became easily offended by things and lost my empathy for others, including my children.

I stopped fishing. What's the point? Life is so short that it can be too short and unpredictable. I started feeling sorry for myself, and my family became resentful that I could not get past it, that I had a second chance, that I was alive, and I should be grateful. I wanted to think and feel these things, and I tried to convince myself, but I struggled.

1.6 CAN'T BREATHE

I was cramped between two people on a flight and, I felt that I couldn't breathe. I have heard stories from family members struggling with travel anxiety, but never me. I usually embraced it, I tuned out, in fact, I enjoyed it. That night I told my wife about my experience and dreaded another flight experience like this.

Has this incident permanently changed me? That was my biggest fear. I only hope that just as my physical scar has healed (and yes, they did a decent job of sewing me up), my mental scars will also fade. I am still seeing a psychologist, which has helped me recognise and adopt strategies to manage my anxiety more effectively, but challenges persist in my daily life.

1.6.1 The Wife's Story: A Story from the Carer and Safety Professional Perspective

At 7:18 p.m. I received a phone call from the surgeon to advise me that my husband's second emergency surgical repair went well. They had removed 10 cm from this bowel and I was told that I could go to the hospital because he was being moved to recovery.

At 9 p.m. they wheeled him into the recovery room, and I was relieved that the emergency surgery was over. I sat next to my husband's bedside, eagerly waiting for his eyes to open. I saw him struggle upon waking as though he were trying to tell me something. His movements were jerky, his breathing was erratic, and he dozed back asleep. He woke again and mumbled, 'I felt everything'. Three words that would change our lives. I felt cold, like a rush came over me. My mind began racing and my gut told me that something was wrong. *Could he be hallucinating? Surely there were still many drugs at play here?* But my gut was telling me otherwise.

He opened his eyes again and asked, 'Where are the kids tonight?' I explained that they were with my parents. He said again, 'I felt everything'. I asked what he meant by that. 'Did you feel tugging or some pulling?' as I recalled my c-section experience of childbirth. He said, 'No, I felt everything, including so much pain. I heard them speaking, and I knew what they were doing. Did they cut me?' I said, 'Yes, they removed 10 cm'. He shook his head. I told him that I would speak to the nurse to understand why he may have believed that he felt everything (I was still doubting this recollection, not wanting it to be an accurate reflection of his experience). At this point, he had fallen back asleep. My anxiety was increasing, and I knew from my gut feeling that something was awry. I immediately went to the nursing station to discuss this with one of the nurses. She advised me that she would contact the anaesthetist right away. I asked whether this was a regular occurrence? 'These occurrences are not typical', she replied, and I could sense some urgency in her voice. I feared now that there was an error in the anaesthetic management provided.

The anaesthetist arrived 30 minutes later, and he looked tired, dishevelled, and muddled with perspiration. By this time, my husband was fully awake, and it was 9:30 p.m. It was an awkward moment, and I started the conversation by stating that my husband said that he could feel everything during the surgery. The doctor admitted that something went wrong during the procedure, and he told me:

> Your husband was awake during some of the surgery. We realised that his vitals became very erratic, and we began troubleshooting as a team. On inspection, we realised that the gas had run out, and he was not asleep during the procedure. We then rectified this and put him to sleep to continue the surgery.

The doctor was apologetic and empathetic, and sat with us for some time because, collectively (practitioner, patient, and family member), we were in shock. I saw this admission as a sign of courage and ownership from the doctor. I am an experienced health and safety representative and a qualified healthcare professional, and I understand issues of clinical and operational governance. As such, I suspected that he would get into trouble for coming to us so soon and openly admitting the failing. I am well too aware that we live in a highly litigious society. As difficult as it was to hear his news, we found that the early admission and ownership assumed by this doctor were beneficial. If there had been defensiveness, denial, or more uncertainty, it would have worsened our experience. Any denial would have made us angry. We have no animosity towards this doctor or the hospital administration. Many friends and family expressed their horror and anger on our behalf, in angst in their belief that we were too lenient and that the doctor must be held responsible and accountable. He may hold some responsibility, but that will not help us. Attributing blame to one person in a system, as though there was a single point of failure in healthcare delivery would not allow the millions of people worldwide exposed to these risks to derive comfort (or us, for that matter), nor would it address the need for sustainable improvements.

When I don my human factors hat, I know that there would have been multiple factors that led to this serious error; essentially, the depletion of the aesthetic gases. *Even when you light your gas barbecue at home, don't you check whether you have enough gas to sustain the barbecue?*

The primary function of a surgical anaesthetist is to provide and maintain a state of deep sleep during invasive procedures. The anaesthetics prevent the patient from feeling pain and discomfort, and a complementary suite of medications stop you from becoming alert and aware of the procedures mid-operation. Through good anaesthetic care, you want pain relief, muscle relaxation, and diminished response to noxious stimuli. However, failed anaesthetic intervention can mean that the medications paralyse your muscles, you cannot move, but you can feel the procedure mid-operation and experience pain from surgery. You can hear voices in the surroundings, remaining in a paralysed state whilst aware, and yet be unable to do or say anything. This was the case for my husband.

1.6.2 Advocating for Improved Health Outcomes

My thoughts began skirting among the roles of wife, carer, health professional, and human factors advocate. So many questions flooded my mind as I began to internally 'investigate' what went wrong? There was a struggle between being the supportive carer to my husband and putting my safety manager hat on while commencing a full-blown investigation leaving no stone unturned. In my mind, I was ready to engage the best external human factors experts to help the team uncover what went wrong through an expert unbiased lens. What went right, too, and what could care teams do better?

I knew that the administration would blame the doctor. However, given my long history of conducting many incident investigations in a range of industries, including logistics, aviation, and manufacturing, I knew that it was never as simple as one person doing something wrong. How can you rely on hypervigilance as your only line of defence in a critical care setting, as needed during invasive healthcare and abdominal surgery?

1.7 THE AFTERMATH

1.7.1 Impacts on the Healthcare Team

I considered the mental health effects on the medical team that night and, being from a health background, I empathised with their dilemma. I ran into the anaesthetist three days later: he waved me down to see how we were doing. He told me that he had a tough time sleeping and, per his dishevelled appearance and darkened bags beneath his eyes, I saw evidence of this. He told me that the three cardinal rules of his profession were to 'keep them alive, keep them asleep, and keep them pain-free'. He told me that the systems needed improvement and that the machines could have been more user-friendly. 'In a complex environment, they can hinder us', he said, 'I have been thinking about this for years'. I did not feel that he was shifting blame; rather, he was just trying to figure things out. He repeatedly expressed regret and said, 'Ultimately, I am responsible for everything that goes on in that room' (shouldering too much of the burden, given that a system of care involves many players and several layers of governance, systems, and technology).

1.7.2 The Hospital's Initial Response

Initially, the hospital contacted us daily. They moved us to the best room. They treated us kindly. However, at times we felt that they were fishing for information and that there was the implication that we were making this up (I cannot imagine anyone wanting to do so!). A part of me thought that the hospital administration was interrogating my to check the validity of his story. Maybe they were hoping that he would forget overnight? There are stories where people forget what happened after anaesthesia during what is known as 'post-operative cognitive dysfunction' (Amiri et al., 2020). I wondered if that is why they waited for us to alert them, rather than initiating the discussion with us and disclosing to us what had happened first.

1.7.3 Multiple Meetings with Hospitals Administration

'Do you have a human factors team involved?' I asked. 'No, but our nursing director has done a short course in human factors'. That was not satisfactory to me, but at least there were some insights into human factors. I wondered about what processes they would use to better uncover what went wrong, and questioned why it went well most of the time? I wanted to see improved system design and understand what care teams, administration, equipment suppliers, and others would learn from this. How could the system improve on this care and prevent this from happening to anyone else and could we trust the system again? The hospital director advised us that their safety manager would lead a full investigation, and that they would notify us within six weeks of the outcome. I would have liked to see an independent investigation completed with the guidance of a human factors' expert.

The hospital followed through with its commitment to meet with us with the investigation findings six weeks later. The finding of the internal investigation revealed multiple points of failure that night that contributed to the incident, one being that the risk of anaesthesia awareness has improved significantly from many years ago with technological enablers amongst other procedure that many of the young nurses and professionals in that room that day were probably not tuned in to the risk of waking mid surgical procedure being a real possibility anymore.

The hospital administration initially committed to a process of collaboration and patient involvement to improve their care delivery. However, we have yet to hear from them again. We do not know if any material changes were implemented.

1.7.4 Learning from Lessons versus Hiding the Experiences and Covering Them Up

How does the healthcare industry move towards a just and fair culture (Frankel et al., 2006) in which speaking up, reporting, and learning team discoveries are openly shared without fear of retribution? How does the legal landscape impact on these ideas? What can the healthcare industry learn from others like aviation, mining, and transport?

1.7.5 Design

The nurse refilled the gas canister mid surgery, she failed to turn it back on. Was there an alarm to alert the machine is off? Is there only an alarm to alert users when it is getting low? I note that both doctor and nurse failed to check that the vaporiser was on. How do surgical teams ensure that they have enough gas for anticipated procedures and what happens when a surgery takes much longer than scheduled? Would better planning have helped?

Doctors and nurses will tell you that operating room machines alarm much too frequently – could this affect the inattention that arises? The care teams told us that the gas alarm was hardly audible, and the alert occurred for a short burst, and quickly became part of the background noise. It would have been difficult to discern and was unlikely to consciously register as a meaningful alert. I wondered: How were health professionals, doctors, and surgical teams engaged in the design of theatre equipment? What forum is there to provide feedback to manufacturers about their equipment design-in-use (Béguin & Rabardel, 2000) to improve critical systems that can lead to fatal consequences? Who holds to account the designers of healthcare systems and the technology adopted in that system? What constitutes a flaw or weakness in operating theatre equipment and in the care delivery systems? Specifically, why do the anaesthesia gas canisters need human intervention to refill them, and is there a better way?

1.7.6 Notifying Adverse Events

It is interesting to note that my husband's experience is not significant enough as a sentinel event to be deemed an adverse notifiable incident to the Australian Commission on Safety and Quality in Health Care [1]. Sentinel events are a subset of adverse patient safety events that are preventable and 'result in serious harm to, or death of, a patient' [2]. They are the most serious incidents reported through state and territory incident reporting systems. However, there are fundamental situations that make this list:

- Do not kill the patient.
- Do not operate on a wrong body part.
- Do not give someone the wrong infant or child.

Does that mean that care teams consider every other incident as an 'acceptable risk'? Do patients know that this is the risk-based threshold tolerance of the health industry and, if they did, would they elect to have surgery for non-life-threatening conditions or routine investigations such as a colonoscopy that, despite their honourable intention to scan for disease states, can lead to complications?

1.7.7 The Impacts on My Family and Me

The trauma and anxiety were extreme for me. For days, I did not sleep. It felt like I was living a nightmare also. After my husband's physical injury healed, there were residual psychological impacts. Even months later, the experience affected our lives – the movies that we could watch, or the alarms that would trigger an emotional response when were out. My husband lost his job, and this took a toll on my health also. I felt aspects of anxiety and panic. Our emotional display upset our children and I cannot quantify the extent of the effects on them.

1.7.8 Now and Where To?

It is one-year post-event, and we need more confidence in the healthcare system. My husband never wants to have to go under general anaesthesia again for any reason. Being an optimist, I believe that everything happens for a reason. If anything, we hope that our story positively impacts our friends and family insomuch as they learn to ask more questions and discuss the scope of care provided by the anaesthetist, surgeon, and other healthcare providers before any procedures. They must ask about how the care team will monitor their health status while they are under general anaesthesia.

Questions remain unanswered for us:

- Are health professionals empowered to speak up or report issues when processes, systems, and equipment operation pose challenges?
- How do contracted medical and health staff interact in operating theatres? What is the level of trust and communication among all parties?
- Do designers and manufacturers of medical and healthcare systems and equipment collaborate with healthcare professionals (the users) to ensure

that the user and patient, where relevant, participate in the design? If so, is this process rigorous and effective?

- Is there an active and continuous improvement process where preventative maintenance and operating teams regularly review and monitor equipment to ensure that it is delivering 'optimal care' (and how is this defined?). What are the opportunities for ongoing enhancement and how is this captured?
- How will the hospital administration share any lessons learned with all the relevant health practitioners and settings if the investigations are closed-door?

We do not know the results of the health system changes or whether the administration made any changes because of my husband's case, but we pray that there are improvements. While our questions remain, we hope that someone is reading this book and is asking the same questions that can influence decision-makers to make necessary changes. We want them to explore good system design from human factors perspectives and improve general anaesthesia monitoring practices. We envision that educational and practice settings can foster these initiatives in equipment design, procurement, supply, training, and use, and that relevant stakeholders can contribute in a collaborative fashion.

REFERENCES

Amiri, A. A., Karvandian, K., Ramezani, N., & Amiri, A. A. (2020). Short-term memory impairment in patients undergoing general anesthesia and its contributing factors. *Saudi Journal of Anaesthesia*, *14*(4), 454–458. https://doi.org/10.4103/sja.SJA_651_19

Béguin, P., & Rabardel, P. (2000). Designing for instrument-mediated activity. *Scandinavian Journal of Information Systems*, *12*(1), 173–190.

Frankel, A. S., Leonard, M. W., & Denham, C. R. (2006). Fair and just culture, team behavior, and leadership engagement: The tools to achieve high reliability. *Health Services Research*, *41*(4 (Part 2)), 1690–1709. https://doi.org/10.1111/j.1475-6773.2006.00572.x

Swannel, C. (2018, September 2). New clinical care standard for colonoscopies. *Medical Journal of Australia*. Retrieved October 1, 2022, from www.mja.com.au/journal/2018/new-clinical-care-standard-colonoscopies

2 'Patients Should Be Seen, Not Heard' *Patient-Absent Care after a Cancer Diagnosis in the Hospital Setting*

Mary

Mary has dedicated this chapter to her two daughters who shine light into her life every day.

At the time of diagnosis, Mary was a sole parent in her late 40s with no substantial healthcare concerns. Three years earlier, a cervical screening test was negative. The latest cervical screening, however, showed abnormalities. Mary's doctor explained that there had been significant changes to screening procedures that resulted in more comprehensive screening. At this stage, Mary and her doctor were unaware of the presence of any cancer and were only investigating abnormalities.

Mary entered the public health system and awaited her first appointment that the hospital administration advised by letter. Unaware that the purpose of the appointment was to perform a biopsy, she came expecting to discuss the screening test. Instead, with no explanation, a male clinician asked her to remove her clothing from the waist down, and to move into the next room to 'that chair' for the biopsy. The clinician appeared more focused on explaining the process to the young male intern in tow than ensuring Mary's comfort. Mary felt mentally unprepared, and experienced a mix of embarrassment, discomfort, and self-consciousness during the procedure. She did not see the clinician again and the nurse advised that her appointment had ended.

At her next appointment, a female clinician launched into explaining the surgical consent paperwork that the hospital administration required Mary to sign. When she asked about her biopsy results, the clinician said in an off-hand way, 'Oh, didn't anybody tell you? You have cervical cancer'. When Mary queried why the doctors had not told her this, the clinician became defensive, claiming that Mary would have received a phone call. When Mary asked for the date that this call took place, the clinician advised that she did not have access to that information. Rather than having empathy for Mary and her understandable shock at this sudden and frightening diagnosis, the clinician insisted that the doctors would have advised Mary. Despite the hospital's claims to be patient-centred, there had been no opportunity for Mary to have a support person at the appointment, no time for debriefing, and no opportunity to ask questions.

DOI: 10.1201/9781032711195-3

Endeavouring to process this devastating news, when Mary inquired further, the clinician used the back of a Patient Information Form (which Mary still has) and drew her a rough sketch of where the cervix is located. When pressed for more detailed information, the clinician said that she was busy, and Mary could ask the hospital staff any questions. She felt pressured to sign the surgical consent forms and clinical explained that she did not have to proceed if she later changed her mind. The clinician shuffled Mary into the hospital hallway to wait, alone, with little context around this news. Mary tried to question the hospital staff, who said that they couldn't answer medical questions, and that she would have to speak to her doctor. Upon reflection, Mary describes this moment and herself, in the context of life before and after cancer, with this experience signifying the bridge between the two.

Mary's next visit to the hospital was 27 days later, for a cone biopsy. Prior to this surgery, Mary held several telephone discussions with the pre-anaesthetic evaluative staff whom she advised of her known allergies, particularly to specific types of medication (non-steroidal anti-inflammatories). There were telephone discussions with nursing staff and clinicians at several levels about these concerns. On the day of surgery, Mary was wearing the standard hospital coloured wrist band identifying that she had allergies of which the surgical team needed to be aware. A clinician entered the room outside of the operating theatre and asked Mary to sign a consent form allowing trainees to be present and participate in the surgery. By then it was late in the afternoon and Mary had been at the hospital all day. Although already under the influence of sedation and minutes from the medical team moving her bedbound body to the surgery, the hospital, nevertheless, required Mary to sign this consent form. She needed glasses to read and did not have them with her, so the clinician pointed out where Mary needed to sign. Looking back, Mary believes this was unethical on the part of the clinician who required written consent on a document which Mary was physically unable to read. This was also unethical in a situation where she was under the influence of medication and feeling vulnerable. Consent forms serve as legal protection for the hospital, and when hospital staff obtain these procedurally but not ethically, they do not serve the patient.

When Mary awoke from surgery, she experienced nausea and excruciating pain in her left flank, for which she medical teams gave her various medications whilst she was in post-Op. It was 10.30 p.m. that night before Mary was stable enough for the doctors to discharge her home. Over the next few days, Mary was continuously nauseous, bleeding, unable to eat, and had a constant sharp pain in her left flank. She saw her doctor who rang the hospital out of concern. A gynaecological staff member at the hospital suggested the use of general antibiotics. Five days after leaving hospital, at 2.00 a.m., with no change in symptoms, and unable to bear the pain any longer, she drove herself 30 minutes to the emergency department. When eventually seen 6 hours later, the clinician on duty burned her cervix with nitrate to stem the bleeding. He also, unexpectedly, mentioned to her that the cancer had spread past the margins of the dissections of her cervix, and that the hospital would be calling her to discuss this at some stage. Again, a nursing clinician gave Mary unexpected, unwelcome news about her diagnosis with no opportunity to discuss the implications. The nursing clinician, albeit kindly, advised Mary that she would have to wait for the hospital to ring her to discuss the matter. She knew not to hold her breath for that phone call.

Mary spent the next few hours in a curtained area, shared with other patients, awaiting an ultrasound to investigate the pain in her left flank. While waiting, nursing staff advised her that samples taken earlier on admission showed that she had an infection from the surgery, and a separate urinary tract infection. Mary processed this information, but a female technician interrupted her thoughts during the ultrasound. The technician, who sat on a round swivelling chair, slipped off and fell onto Mary, so that her bodyweight and the scanner pushed deeply into Mary's pelvic area. The technician, though embarrassed, remarked awkwardly to her colleague that this was the first time that she had fallen on a patient. They quickly moved Mary back to the curtained area, where her vaginal bleeding started again.

Late that evening, the orderlies moved Mary to a bed on the ward. She had been so ill when getting herself to hospital the day before that she had not thought to pack a hospital bag. So, she searched the ward until she found supplies, and, importantly, her pads because her bleeding was now constant. Mary had to be resourceful to procure supplies considering the reminder she received from staff that she was not vacationing in a motel.

Around lunch time the following day, a nurse told Mary that the hospital was discharging her. A minute or so later, a female clinician arrived to send her home. An orderly arrived with a wheelchair to take her to the discharge lounge, where she waited over an hour, too unwell to drive herself home, though drive she did. For Mary, it was a slow and lonely drive home, her head spinning from the events at the hospital and the news that the earlier surgery had not got all the cancer. Once home, the bleeding continued for several days, stopped, then started again. The left flank pain continued, with Mary's doctor unable to offer any explanation.

She received the letter setting out her next appointment with the hospital gynaecological team and tried to make an appointment with a private gynaecologist for a second opinion, but the earliest possible date was after this hospital appointment. At the next hospital visit, a different female clinician physically examined her to measure whether there was enough cervix left to do a second cone biopsy. Mary said that rather than another cone biopsy, she wished to have a hysterectomy. When pressed about a hysterectomy, the clinician remarked that that was not the way they do things in the public hospital, and the next step would be to undertake another cone biopsy. Mary certainly could not afford a private surgeon or private hospital, so she believed that she had little choice.

Mary had an appointment the following day with a private gynaecologist. After showing him the results of her histopathology reports and discharge summaries, he advised her that if she were his patient, he would recommend a hysterectomy. This clinical reasoning conferred with Mary's ideas. However, at the end of the appointment, the gynaecologist asked the name of the head doctor of her hospital team. When Mary told him, the specialist changed his tune and said, 'Well if (name of doctor) has said to do another cone biopsy, then that is what you should do. He knows you best, so you need to go with his advice'. Mary thought: *He knows me best? How can he, he has never met me, never spoken to me, and not a single member on his team has listened to a word I have said.*

Exhausted and unwell, the next few weeks were a blur. The doctors performed a second cone biopsy. Again, post-surgery Mary awoke to the usual nausea, severe left flank pain, and an inability to pass urine. It was past 11.00 p.m. at night but, again, despite feeling very unwell, the doctors discharged her home. Her discharge notes stated the surgery had been an 'uncomplicated but difficult procedure'. Four days later, too ill to drive, Mary called a friend and asked her to, once again, take her to the emergency department. After another few hours in the emergency department, the medical teams admitted her to the day ward, experiencing acute kidney failure. After 24 hours on a drip to restore her kidney function her levels had dropped, and the staff advised Mary that they would admit her to the general ward.

Throughout the stay, she heard constant bickering, often outside her door where the main desk was situated, among the renal team, the head of general ward, and the gynaecological team. Each team attempted to distance themselves from any responsibility for Mary's care, and from any accountability for her kidney failure. There seemed to be no communication among different teams as they organised different scans or biopsies without the others' knowledge, which the ward doctor cancelled. On one occasion, no meals were served to Mary whilst the teams argued over the need for a renal biopsy. The doctors cancelled the procedure and Mary, ill and light-headed from a lack of food, received only dinner that day.

On day three, the ward doctor advised her that the kidney readings had improved and that she could go home. She still felt terribly ill, and the pain in her left flank had not improved. She asked repeatedly for details of what her kidney readings were, however, the doctor refused on every occasion to discuss or provide these details. Mary made this request to any clinician who walked in her room, and all stated they would ask the doctor and get back to her – but none ever did. She contacted patient liaison services to ask for support.

A meeting was organised for the next day but about an hour before the meeting the ward doctor appeared, stating that he had no knowledge of any meeting and would not be available at the arranged time. Again, the doctor refused to let Mary know her kidney readings. She organised for a support person to attend this meeting with her because a patient liaison officer was not available to attend. The doctor returned at that time, again refusing to discuss her readings. The ward doctor continued to be unwilling to provide any details of Mary's situation, nor was he open to any discussion around the cause of Mary's kidney failure. Finally, he said that he would arrange for his assistant to provide Mary with the kidney readings. Unsurprisingly, the assistant never returned with any information.

Mary began to suspect that a medical error was behind her kidney failure and she asked if she could see the hospital pharmacist to gain access to details of the medication that the hospital had given her during her surgeries and her hospital readmissions. The care team advised Mary that the pharmacist would not be available to meet with her. Why was her medical information kept as a tightly guarded secret? She found it ironic that each team seemed defensive if not informed about certain tests or scans being organised or authorised without their knowledge. Yet, none of the clinicians were willing to provide Mary with *any* information that related to her health status.

Patient Liaison Services stepped away; they had organised the meeting, and from their perspective, they had completed their work. Mary's health was not improving, and her mental health was deteriorating. There had been a brief visit from one member of the gynaecological team to advise that they had not fully excised her cancer in the second operation, and it was now outside the margins of the cervix. The team member advised that she would need a full hysterectomy. Overwhelmed, Mary asked to see a social worker, which on reflection was a big mistake. Now there was a real push to get her out of the hospital. Again, no social workers were available in the hospital and the hospital medical teams discharged Mary the following day.

On the afternoon before discharge, one kindly nurse, who had witnessed the daily debacles and Mary's deteriorating well-being, spoke to her privately. She said that the team believed that her adverse reaction to medication during surgery had caused her acute kidney failure. The nurse offered to advocate for the pharmacist to provide her with a medication list. As a result of this nurse's intervention, the pharmacist gave Mary a list of her medications before she left hospital. Stunned, she immediately identified that the medical team administered non-steroidal anti-inflammatories despite her efforts to advise them verbally and in writing that she was allergic to them. She had even worn the coloured wrist band into surgery that identified that she had allergies. She had informed the anaesthetist outside the surgery, and his team in the weeks leading up to surgery, that she was allergic to non-steroidal anti-inflammatories. Nevertheless, the medical team gave her anti-inflammatory medication during her second surgery. She was devastated that not one clinician had listened. And now there were other related complications, because Mary's extended hospital stay resulted in a blood clot in her leg. The medical providers monitored this via ultrasounds rather than medication because of her (finally recognised) allergies.

The 'About your Anaesthetic' brochure stated in part, 'Your anaesthetist is a doctor with specialist training who will . . . agree to a plan with you for your anaesthetic and pain control'. Despite being clear that the patient was allergic to non-steroidal anti-inflammatories, the anaesthetist did not follow the agreed patient care plan. The context for this agreement had been based on the information on the patient records, the patient's wishes, and the allergy alert wrist band to exclude non-steroidal anti-inflammatories in their anaesthetic medication. The anaesthetist did not advise the patient that they would not follow the agreed plan.

The surgical support staff scheduled Mary's hysterectomy for two months' time. In addition to the usual appointments with the pre-anaesthetic staff, and due to her compounding complex history arising from the medical mistakes, she scheduled appointments with the head gynaecologist and anaesthetist. When she arrived for her face-to-face pre-anaesthetic evaluation, the clinician was defensive. Mary sought an explanation as to why, despite advising that she was allergic to non-steroidal anti-inflammatories, the anaesthetist included these during her surgery. The anaesthetist contradicted the information that medical teams provided to Mary earlier about her hysterectomy, claiming that the surgeon would remove Mary's ovaries. She did not expect this, although she was unsurprised by the conflicting information: the right hand did not know what the left hand was doing.

The clinician was not prepared to acknowledge the distress that the medication error had caused Mary. She mentioned the recent blood clot that arose during her last eight-day hospital stay. Neither this clinician nor his senior supervisor could decide how to manage this. They determined that she would now need to take, via self-injection twice a day, medication to clear the blood clot. They sent her to the hospital pharmacy where she spent the next hour waiting while the pharmacist and clinician discussed on the phone, back and forth, her suitable dosage.

Next, she had an appointment with the head gynaecologist. Throughout the appointment he spent more time concerned by the printer, which was not working, than focusing on Mary and discussing her upcoming surgery. The gynaecologist told her that they would have to cut her open at the abdomen for her hysterectomy due to the elevated risk of damage to her kidneys. Had she had the hysterectomy when she first requested it, Mary would have had a choice between this and a vaginal hysterectomy. Now she would have a scar across her abdomen for the rest of her life to remind her of their ineptitude.

The doctor advised that if she had asked for a hysterectomy previously, she could have had one; that the hospital would have respected her decision. When she mentioned that she *had* requested one from the start, the doctor dismissed her protests. She asked if she might receive support from a following surgery, knowing that there were many obstacles and hurdles that she would be facing in her post-surgical care. The doctor advised that there were no social workers who were available during her upcoming hospitalisation period. Mary dutifully signed the consent forms; however, she did not consent to allow students to operate on her during surgery. Mary drew a line through this part of the consent form and the doctor quickly ushered her out the door: her choice was an unpopular one. So, she injected herself morning and night and her stomach was now red and covered in bruises from this painful twice-daily process. Meanwhile, her vaginal bleeding continued.

On the day of surgery, whilst the surgical team prepped Mary, another clinician came in with a fresh set of consent forms to re-sign. The head doctor had inadvertently completed the forms and filled in his name in the field nominated for the patient, so she needed to complete new forms. No surprises there! Again, Mary refused to sign a consent form stating that students could conduct surgery on her, and the clinician left to seek further advice. As with the previous surgeries, whilst the anaesthetist sedated Mary in the pre-operative room, a student appeared asking her again to sign the consent form. Again, Mary stated that she was happy for the student to observe; however, given all the medical mistakes she had experienced along the way, she would not consent to students taking part in the actual surgery. With no glasses, a form she could not read, and going under the anaesthetic, a student again directed her to sign the consent form before the surgery could proceed. Mary will never know if the hospital allowed students to operate on her in this surgery.

Whilst recovering on the ward, there were few discussions with Mary about her post-operative care. One day, Mary's urine bag was so full, it backed up and she was in severe pain, having called for the nurse numerous times. The pain became excruciating, and Mary's daughter sought help. After another half an hour, a nurse came to find the bag nearly bursting. Mary sometimes woke to find some brochures left by her table, or a clinician came in and said that they were too busy to talk, but there were some brochures to read. Once more, there was no opportunity to take in information, process it, or ask questions. They had done their job and could tick the box: informed patient about x, y, or z. For Mary, given the brief time between diagnosis, three surgeries, numerous medical mistakes, re-hospitalisations, and ill health, she felt as if she had been sucked in and then spat out the other side of the system. There was no time to catch her breath, seek help, or process what was happening. Looking back over the journey, the only positive for Mary is that after finally having undergone the hysterectomy, the cervical cancer is now gone. However, her body carries the visible scar across her abdomen and the discomfort of the left flank pain that continues to this day. Throughout the journey, apart from having cancer and normal post-operative recovery, Mary experienced acute pain, reduced mobility, fatigue, nausea, bleeding, acute kidney failure, low concentration and mood, adverse drug reaction, infections, diarrhoea, and blood clots that impacted her ability to function and to provide for herself.

Mary wears the emotional scars from her poor healthcare experiences, and the grief and trauma of the physical ordeals. Mary was torn, unable to work for much of this time and, at times, humbly living out of a rented room, a confronting experience for a professional woman. Two months after the hysterectomy, Mary moved interstate to start a new job. She emerged from an experience that, although consumed only six months of her life, it felt more like six years. Since then, Mary also had a colonoscopy/biopsy, which she had been on a long wait list for, to check for early colon cancer. The doctors gave Mary the 'all clear' and the doctors performed the procedure at a different hospital. However, again despite advising staff and writing on all hospital admissions forms her history and allergies, hospital staff gave Mary a white wrist band which signified that she had no allergies. After spending time getting this rectified pre-admission, again when Mary was waiting in the room outside the surgery, and now wearing a coloured wrist band, the anaesthetist was yet still unaware of her allergies. Mary waived her coloured wrist band in front of him and another discussion ensued minutes from surgery. There were no surprises to Mary when the discharge summary for the procedure similarly reflected the hospital system's inability to keep accurate records, noting Mary's allergies as 'nil'.

From these experiences, Mary believes that the narrow gaze of the medical model does not allow for a holistic view of the person, their context, their needs, or their rights.

RECOMMENDATIONS:

- Healthcare administration must change the structural and systemic restraints imposed on doctors by their discharge pressures that overrule patient-centred care and recovery.
- A patient should have a right to information related to their own healthcare. Hospitals need to establish processes that enable patients to have access to their own information. The failure of hospitals in being transparent about patient care inhibits their capacity to provide safe, person-centred care.

- Patients need to be involved in decision-making and planning about their healthcare.
- Hospitals should have ethical consent processes in place that do not require a patient to sign consent forms when a patient is physically unable to read the document.
- Healthcare ethics and administration teams must establish ethical guidelines that afford patients the opportunity to consider consent forms when the patient can make clear decisions - outside of when they are going under sedation and minutes from the actual surgery. If a patient refuses consent to students operating on them, then hospital care teams must respect the patient's right to refuse.
- Hospitals should refer to a patient's history, their documentation of medication allergies and adverse drug reactions, and the allergy alert wrist band that they wear to minimise the risk of adverse drug reactions.
- The hospital quality care team investigations of the patient experience of an adverse drug reaction should include their consultation with the patient.
- Consumer complaint teams/patient liaison officers/social workers are poorly resourced in the hospital setting. This may speak to staffing issues and value placed on funding for these services. These roles need to be better resourced so that they are accessible to patients.
- The quality care teams of hospitals must embed patient-centred care frameworks in their practice – this includes respect, emotional support, physical comfort, information and communication, continuity, and transition, care coordination, and access to care.

3 Reflections on 'Patients Should Be Seen and Not Heard'

Sara Pazell

I have read and re-read this chapter many times. Each time, Mary's experience of trauma sways me. Astoundingly, there were opportunities for improvement in every interaction with those whose duty it was to provide her with care and protect her health. I see this as the system of care failing her, not any one provider. Yes, people may have been rude and shirked their duties, and the cost of quality and efficacious care was significant, but how did the healthcare system provide for those delivering that care? Was there adequate time in the schedules of the providers to ensure quality and thoroughness in their procedures? Could unhurried conversations foster better relationships? Did the environment inspire hope, health, and creativity among all who shared that space? What psychological safety protections and health-promoting activities existed for those who must care (the workers) for others (the consumers) during these interactions? It is recognised that workers contract their generosity, volition, and interest in their service delivery during times of psychological burn-out (Maslach & Leiter, 2016). Emotional energy can be depleted, indifference may reign, and self-efficacy can diminish (Edú-Valsania et al., 2022). If a healthcare worker no longer cares about themselves nor feels effective in the care that they provide (perhaps because the effort becomes awash and diminished in a poorly designed system), how can they provide inspirational, loving, or empathetic service to another? At the very least, the care delivery should be competent.

What confronts me in this story, among many facets, is that Mary is a highly educated and forthright woman who can hold her own. Yet, even in this relative state of privilege, she is powerless. If this is the case, what chance does someone with a cultural, language, literacy, or disability barrier have when they must advocate for their care in a broken or, at best, struggling system? Healthcare presents its users with an imperfect system: for example, simple medication prescription errors are the norm rather than the exception. In a study of four Australian and New Zealand hospitals, from a sample of 200 drug charts, an error was found in 672 of 715 patients (94%) (Barton et al., 2012). This is remarkable. Studies in the leading practices to reduce anaesthetic medication errors are emerging and suggest that this must occur through the partnership of a pharmacist–anaesthesiologist team (Renaudin et al., 2020).

DOI: 10.1201/9781032711195-4

To exemplify Mary's story, I depicted key interactions and service errors on a journey map (Figure 3.1). It is harrowing to know that such dramatic and confusing circumstances can occur in mature healthcare systems. The journey map provides a visual storyboard (Maguire, 2001; Rosenbaum et al., 2017) to add meaning and interpretive value to the experiences. In Mary's case, this meant six months of trauma with three surgeries, two emergency room visits, and two re-hospitalisations, each resulting in complications. Notwithstanding, there are the psychological and physical residual effects of the procedures and the interactions that Mary must deal with now.

Cover-ups, mishaps, and poor system design cloud what might otherwise cause the fulfilment of an ethical medical ethos to do good. Patient lives are in standstill during traumatic health events. They may not accept themselves as a cog in someone else's machinations of work when the impact on their life is significant: life changing, in fact. Nor should they. Yet the administration, the public, or the legal fraternity must not blame a worker for these machinations if system design could improve their performance.

Healthcare systems are dynamic and complex yet afford such rewarding opportunity to do things better. In Mary's case, we owe this to her. The purpose of this book is to shed light on the ways that we get things wrong, how we might get it right more of the time, and how we construct new pathways of care, fuelled by design strategy. Through the consumer stories in the first part of this book, the authors encourage empathy to help inform design improvements.

In a human-centred design world (International Standard Organisations, 2016; Maguire, 2001), we might scope and plan for usability, determine the goals, and understand the context of activities that occur. We must study the science and the empirical evidence, engage the users in the system, and understand what people must do to meet their performance objectives. This involves understanding the technology, environment, and motivators; the way that work strategists design jobs and tasks; and the persuasions of the organisation, such as how they communicate and celebrate design. Empathy is central to this practice and can be depicted by case studies, personas, journey maps, and visual storytelling. Modelling, prototyping, iterative design, and evaluation tools are essential to effective design processes. Design philosophies, concepts, and strategies must evolve through their iterations, trials, and evaluations. As strategists implement and evaluate work design, they must celebrate and communicate their successes. In Mary's case, every touch point in her care could give rise to such analysis and redesign: her medical appointments, the diagnostics, how the doctors determined her surgical needs, her pre- and post-operative care, and the medication errors. Let us hope that these stories inspire ongoing investigations and design improvements.

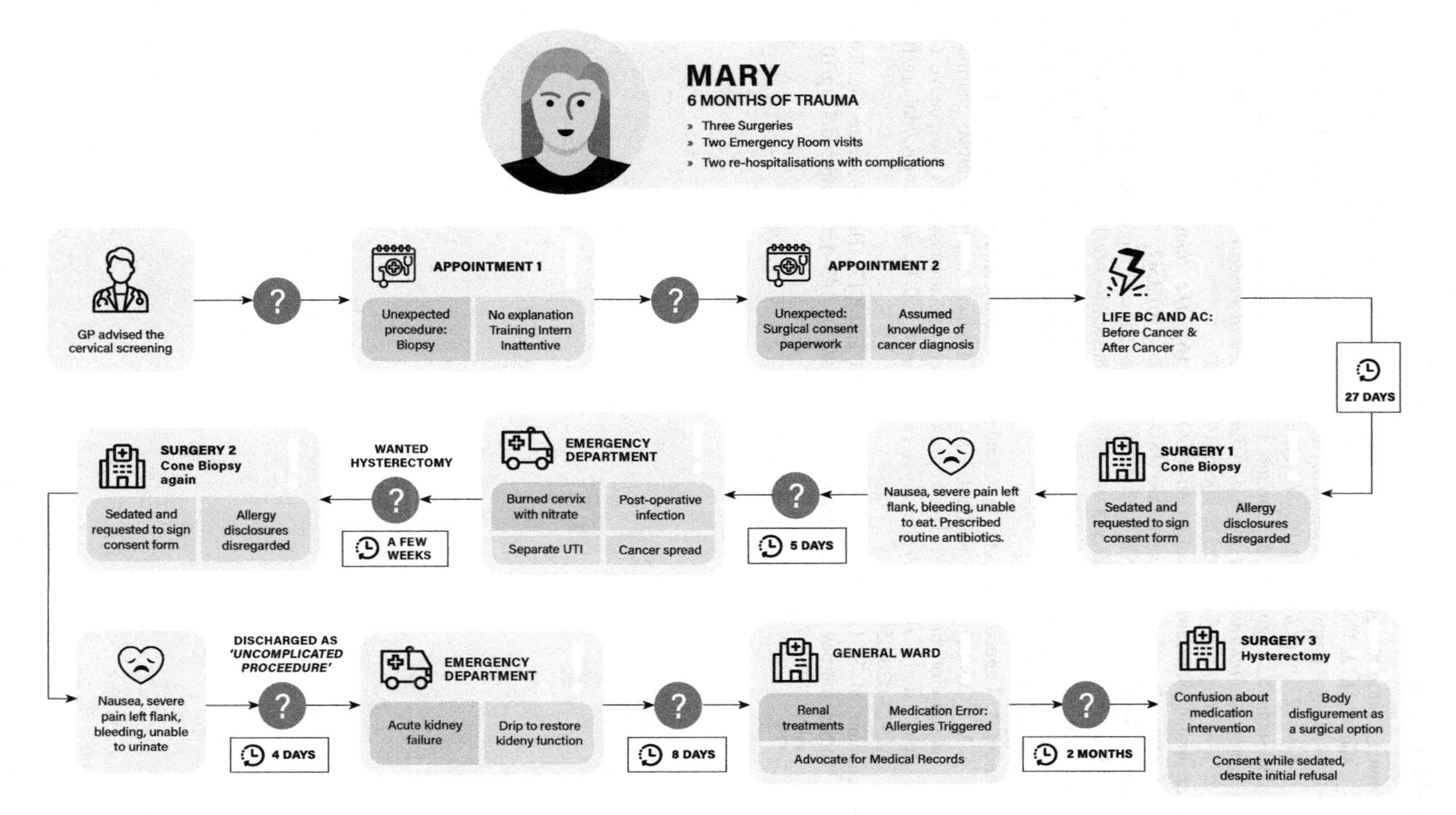

FIGURE 3.1 Mary's six months of trauma.

REFERENCES

Barton, L., Futtermenger, J., Gaddi, Y., Kang, A., Rivers, J., Spriggs, D., Jenkins, P. F., Thompson, C. H., & Thomas, J. S. (2012). Simple prescribing errors and allergy documentation in medical hospital admissions in Australia and New Zealand. *Clinical Medicine*, *12*(2), 119–123. https://doi.org/10.7861/clinmedicine.12-2-119

Edú-Valsania, S., Laguía, A., & Moriano, J. A. (2022). Burnout: A review of theory and measurement. *International Journal of Environmental Research and Public Health*, *19*(3), 1780. https://doi.org/10.3390/ijerph19031780

International Standard Organisation (ISO). (2016). *The human-centred organization – rationale and general principles* (ISO 27500: 2016). ISO.

Maguire, M. (2001). Methods to support human-centred design. *International Journal of Human-Computer Studies*, *55*(4), 587–634. https://doi.org/10.1006/ijhc.2001.0503

Maslach, C., & Leiter, M. P. (2016). Understanding the burnout experience: Recent research and its implications for psychiatry. *World Psychiatry*, *15*(2), 103–111. https://doi.org/10.1002/wps.20311

Renaudin, A., Leguelinel-Blache, G., Choukroun, C., Lefauconnier, A., Boisson, C., Kinowski, J.-M., Cuvillon, P., & Richard, H. (2020). Impact of a preoperative pharmaceutical consultation in scheduled orthopedic surgery on admission: A prospective observational study. *BMC Health Services Research*, *20*(1), 747. https://doi.org/10.1186/s12913-020-05623-6

Rosenbaum, M. S., Otalora, M. L., & Ramírez, G. C. (2017). How to create a realistic customer journey map. *Business Horizons*, *60*(1), 143–150. https://doi.org/10.1016/j.bushor.2016.09.010

4 A Patient and Family Journey through Systems of Care

Carlo Caponecchia

4.1 INTRODUCTION AND BACKGROUND

My dad was admitted to hospital for a period of three months in late 2020 through early 2021 due to an infection which resulted in septic shock. His three months in private hospitals comprised a harrowing 17 days in intensive care, followed by time in several wards, and in a rehabilitation ward of another private hospital, before a further three months of home-based rehabilitation. At the time, he was in his early 80s, spoke English well (though as his second language after migrating to Australia in the 1960s), and had industrial deafness (untreated at the time). His general health before this hospitalisation was quite good, though he had ongoing arthritis, particularly affecting his knees and shoulders, and had surgery years before for a torn muscle in his shoulder which never regained its full function. Around seven years earlier, he had been in hospital with sepsis and, accordingly, his ongoing care needs included wound monitoring and control. Following a childhood of demanding work on the family farm, and then in a restaurant in Rome, he worked his whole adult life in construction in Australia – concrete cutting, drilling, and finishing – including, for a significant period, as a sole trader.

As a human factors researcher with a wide range of experience across different domains (e.g., health, transport, and construction) and expertise in workplace safety, safe systems of work, and particularly psychosocial risks (see Caponecchia & Wyatt, 2011; Caponecchia & Wyatt, 2021; Caponecchia, 2019), this experience raised many interesting (and distressing) human factors and ergonomics (HFE) issues. Some of these I tried to address as best as possible at the time, and others remain in need of documentation and discussion. This case study may help identify risks and opportunities for HFE-based interventions in systems, and could be useful in the wider scope of healthcare, aged care, and rehabilitation.

4.2 INTENSIVE CARE

Getting the right information, which is sometimes rapidly changing, was crucial to coping with the situation. Fortunately, communication at this stage was quite good. We had access to doctors and amazing nurses who could answer all our questions. And I asked lots of questions. I recall one exchange with the doctor when we discussed dad's status. Instead of just telling us what was happening, he asked for our

DOI: 10.1201/9781032711195-5

understanding first. Once asked, I was able to state in detail what I knew about dad's current state – from the functioning of his kidneys, heart, and breathing capabilities, to what signs of development or deterioration we were looking for to inform the plan for the next stages of his care. Given this opportunity to summarise what I knew, the doctor could determine my knowledge base. He then added, adjusted, or elaborated on my understanding. On reflection, this was an important exchange: the doctor sought our understanding on matters and valued our contributions. The doctor evaluated our family's knowledge of my dad's condition so that he could refine this and add granular detail informed by medical expertise. We were part of the team. I also realised that I have privilege here – I am educated and deal with communicating complex information every day. Not everyone has those skills, and I couldn't help but wonder what would happen to someone who didn't have those skills or didn't have someone to advocate for them.

In the following days, we did 'head-to-toe' reviews with the on-duty doctors and the nurses so that we could better understand what to look for, what new treatments were occurring (e.g., dialysis), and how my father progressed. Never have I been more in awe of nurses. These specialists were able to contextualise for us what it is like to be in intensive care for so long, and to actively engage us in dad's care. He was confused and lost muscle tone, which was scary, but they reassured us that this was normal during long stints in intensive care. He needed constant reassurance of his progress, encouragement, and the required assistance to eat and walk. I believe the social distractions that our family visits provided were also helpful. His attitude was negative, and he was confused and sleep deprived. We, his family, were also sleep deprived and not looking after our own basic needs. Dad being in hospital was not the only thing going on. For example, the car battery went flat when we were at the hospital, and that silly, objectively quite unimportant, and easily-fixed thing almost caused our complete mental meltdowns. I know from my practice, reviewing and opining on cases of psychological injury in workplaces, that there is good reason the straw on the camel's back is proverbial. It was a time of extreme stress all round, which is important to remember when considering the other safety and HFE issues that occurred next.

> No one is meant to spend 17 days in intensive care. It is, as the name says, intensive. In addition to the presence of life-saving equipment . . . the space has no natural light, can be quite noisy, and, as a result, sleep is elusive. It is as disorienting for patients as it is concerning for their family.

4.2.1 The Wards

Moving out of intensive care was a significant milestone, but several new challenges arose. The experience of the three different wards was necessarily different to intensive care: it moved from intensive one-on-one monitoring to a different staff-to-patient ratio, and the requirements of all the stakeholders (doctors, nursing staff, allied health, patient, family/advocates) were different. Dad had lingering distress and confusion and increasing dissatisfaction with the length of his stay in hospital, coupled with unclear information about why this was taking so long. The days must have blurred for him – he spent each day staring at the same walls. Walking and exercise, and physio when possible during weekdays was an important highlight because it was a chance to move forward. But nights remained long, painful, and disorienting.

4.2.1.1 Calling the 'Safety Hotline'

There was an incident that occurred during the night that caused dad great concern and pain, and we could not get a full account of what occurred. We couldn't fully understand what happened during the night, but we knew enough to be concerned. Dad reported that one of his lines dislodged (or he pulled it out), and that there were some heated exchanges with nurses regarding how, why, and by whom the line(s) were removed. There was a suggestion that misconduct or, at least, poor practice occurred.

I had seen the posters all over the hospital advertising the 'safety hotline' with a number that that consumers could call should patients or family members be concerned about patient safety. This is like hotlines used in many other industries to enable people who are external to the organisation to report safety concerns. For example, there are often signs on the back of trucks stating that 'We value safe driving. If you see anything of concern, please call . . .' The programme is based on the REACH programme from the New South Wales Clinical Excellence Commission,[1] and systems like this are common. Weeks earlier, when I first saw the sign for the safety hotline, I sensed that at some point I would have to make a call. Unfortunately, I did.

Calling the line escalated the issues and meant that I got access to a more detailed review of dad's condition with the nursing manager, who communicated further with the Nurse Unit Manager (NUM). I received a return call with a full verbal account of dad's file, but no meaningful information about the event that caused me to call. Nonetheless, the complete information I received about his condition was helpful. I used the structure of that 'complete review' as the basis of every subsequent phone call with the NUM so that we had a comprehensive picture of what was going on, and so that I could help answer dad's questions (to me) about his condition and help reassure him. In hindsight, they surely became sick of me calling and appearing to grill them, but this was information that he, the patient, was asking for, and asking me to access. There was an element of 'translation' here, not across language, but across information complexity, and it may not have been medically accurate by the time it got to him. But that didn't matter: it was important for him to have as much information as possible to reassure him of progress and what to watch out for, and what we, and he, could do to speed his recovery. It was about regaining control of a very uncontrollable situation, and we know that control reduces stress.

When I spoke to dad the day after I called the hotline, he asked me what I had done overnight because his hospital experience had changed. He said that he'd never seen so many nurses in his room or had so much attention and people checking on him. Increased activity was not the intention of my use of the hotline. I didn't want extra attention on him; rather, I wanted information on exactly what happened during

[1] www.cec.health.nsw.gov.au/keep-patients-safe/reach

the night shift. Whatever it was, it distressed him and made him distrust everyone involved in his care provision. There was increased activity and a sense that we were watching and ready to report poor behaviours. Overall, that soured the environment, and it is something that concerns me for overall patient care. I didn't want anyone to get into trouble, and reporting a safety concern shouldn't lead to punishment, threats, or even surveillance, but rather support and learning.

4.2.1.2 Equipment Provision and Daily Tasks

One of the mixed messages we received was about the supply of continence pads. Some staff told us that we needed to buy these and supply them ourselves, while others said it was not necessary. This was frustrating because between hospital visits, we scrambled to locate, buy, and collect the right supplies during the COVID-19 supply shortage. One positive thing that we were able to do was in relation to dad's clothes. As he improved, he needed a range of comfortable clothes that were suitable for going to physio sessions, walking around the ward, and in which he could be comfortable sitting in all day given there were various tubes and dressings attached to him. We bought the clothes, washed them each day, and returned them as complete outfits in zip-lock bags, so that each bag contained everything he needed to wear for the day. Nursing staff could just grab a bag and know that they had everything needed, without having to search for random clothes. We had feedback from staff during his stay that this was a great idea, and that it really saved their time. We were happy to do *anything* that could save nurses' time. It was a simple solution, and I wondered why this was not a standard practice that all families or carers understood (if patients have family available to provide such assistance).

4.2.2 Assessment and Uncertainty

The interface with other hospitals and healthcare providers, including the provision of the home care package, was bumpy. The length of dad's hospital stay was caused by the dressing that was used on his main wound – a vacuum-assist closure of a wound (VAC) dressing.[2] I didn't really know what this was, though it was an important, more effective dressing than conventional dressings and, while effective, it also delayed home-based care. The providers of the home-based package would not take someone with a VAC dressing because of the care implications, and there was considerable uncertainty over what to do to continue his treatment. An affiliated hospital was going to assess the case, and initially they said that they would not take him. About a week later, they said that they would. I should have 'Googled' what a VAC dressing was so that I could explain it, but in the myriad of the things that we were managing, I did not think about it. As the key issue holding up transfer (home or to somewhere else), what the dressing was, how it worked, and why its benefits outweighed the costs (not going home), the care providers should better explained this

[2] www.hopkinsmedicine.org/health/treatment-tests-and-therapies/vacuumassisted-closure-of-a-wound

concern. Again, this could have increased that sense of control and understanding about why things were taking so long.

The care teams eventually transferred dad to the rehabilitation ward of another (affiliated) hospital late on a Friday afternoon. The hospital did not alert us: we had no information that this was going to happen, nor did we know about the timing of it. I recall a series of particularly distressed phone calls, when dad did not know where he was, and he was concerned about the new hospital knowing the story of what had happened, and whether they had his medication or not. He is always in control of his medication, so this may have been something playing on his mind (and serves as another exemplar of a lack of control). I had to call the ward to broker reassurance, but really did not get any information about what had happened. I was out of town at the time, and remember having visions of dark, abandoned, echoing hospital corridors. That is a bit dramatic, but this was a low point, and it was extremely distressing, in part due to my remoteness. I had not considered it before, but not knowing what happened in the incident where dad's lines were unattached, for which I called the hotline, played into my concerns here. Dad's trust in the system had eroded; and mine was fast diminishing, as I could not get a direct answer, and now hospital team acted without warning or consulting our family or dad. There should have been handover with reassurance for the patient and his advocates, in addition to a medical handover. Of course, they were itching to get rid of dad from the first hospital: he had been in intensive care for over two weeks, and in the wards for several more, and we knew that the facilities did not want to take him with the VAC dressing (for some unexplained reason). Nonetheless, they could have professionally managed these events. Instead, the hospital teams left us in a holding pattern over the weekend to later learn what the new ward would be like. A weekend of waiting is a long time when in hospital.

In the rehabilitation ward, there was a social worker who we could contact to help arrange a visit to align with the on-site schedule of the treating doctor. This was a great development. I changed my schedule to be present at rounds (which I do not think the medical staff appreciate but, by this point, 'Whatever'). There was a reluctance to discuss any issues without the patient, or without directly addressing the patient, which I absolutely understand. This is an important principle, with patient-focused care. As advocates, we were an important part of the communication chain to translate the information into real terms, reminding dad of the information later (there was a lot of information and no way for him to access or check it after the doctor left). Also, we ensured that we presented the questions that he had, that the care teams answered these, and that dad and we understood the information. There is, of course, a demand characteristic when a doctor asks a patient if they have any questions. Dad had loads of questions, but no one was able to elicit them, or create a space/time in which he could ask them. The doctors and nurses did not consider basic human factors despite implementing a well-intentioned, ethical practice. So, I needed to be there to ask the questions that he had already asked me, respectfully of course, and with genuine appreciation for the efforts of the staff. I also needed to have information ready when dad inevitably had questions later. In this way, care functions that facilitated the understanding of the care process and health status to family/advocates were just as important as talking with the patient.

Factor in language, poor hearing, disrupted sleep, disorientation, and mild situational depression coupled with the complexity of information that required translation, and I wondered how we could expect to have meaningful communication (and comprehension) in this 'patient-centred' practice.

4.2.3 COVID-19 Protocols

All these things happened during early 2021 with heightened COVID-19 precautions. There were limits on the number of people who could visit hospitals and masks were mandatory. It was meant to be two people per day (and the same people), but we received written exemptions while dad was in intensive care. This was important, because during his hospital stay, dad needed visitors twice a day to keep his spirits up and to help communicate about his progress. The COVID-19 restrictions did not make sense to him, and not having family there emphasised a sense of abandonment, coupled with an endless disorienting and depressing environment. Dad did not know if he was going to get out of there. Organising who could see him was a stressor for the family too. Another hospital, to which medical teams and administration almost transferred him to, had a one visitor per day limit, with no swaps (i.e., it had to be the same person) and this was only from 1 p.m. to 2 p.m. It would have been almost impossible to advocate for him in-person under this arrangement, and we had to be able to advocate. The COVID-19 protocols were important, and sometimes the hospital militantly enforced these, and yet at other times they did not enforce them at all (e.g., after office hours), so how important were they really? There were major holes in the proverbial COVID-19 protocol Swiss cheese, which undermined any infection control strategy and just increased the stress of all the players.

Other COVID-19 protocols were alarming for ergonomic and inclusivity reasons. My mum has limited dexterity in her fingers and her familiarity with using a smartphone was new at the time (because everyone needed one for QR-code COVID-19 check-ins at all public locations). The questionnaire attached to the hospital QR code presented 15 questions to answer, in small font, which the user had to manipulate on screen each time they signed in to read the text, and then manipulate again to locate the radial button and answer the question. It was extremely difficult to use for someone with limited finger dexterity and this added to her distress. I could complete it for her – no problems – but redesigning the interface to be user-friendly and still informative is part of facilitating the independence of older people (or people with disability, or whatever user needs there might be). It was frustrating enough to use for people who didn't have these limitations, and so the hospital could apply universal design principles. In addition to being frustrating, causing embarrassment, and being time-consuming, the outcome of this is that it nudged people to avoid completing the questions, which compromised the health of everyone in that hospital. This was an instance of how human-centred design could improve performance and achieve the core objectives of disclosures, compliance, and fostering a community of health and well-being. The obvious retort here is that the health system was under duress and short staffed, which is true – but this was an accessibility issue, controlled by an IT department. An integrated systems view of what the hospital does, with accessibility, health, and well-being across all business units might have resolved this problem.

4.2.4 Invoices

Throughout the three months in hospital, and into the time of rehabilitation at home, we received an avalanche of invoices in the mail from the imaging company. It was quite a significant administrative load, partly because we did not recognise various doctors' names. I had to develop a system for checking and tracking what I had paid. The company sent duplicates if the originals were overdue, but they were only overdue as we were waiting for another letter with a cheque from another agency (Medicare). It was an extremely complex process. This was stressful on top of the twice-daily hospital visits and all the other tasks that we were managing in support of dad's care, even though I had set up a spreadsheet to help me organise this.

I was able to authorise direct payment by the private health insurer, so we only had some invoices for which we had to wait for the public health scheme (Medicare) to send cheques which had to be mailed back to the imaging company. This seems like too much detail here, but that's how complex the process was: we received mail containing sometimes unnecessary and/or duplicate invoices when some could be ignored, and others required attention. We might have received a cheque from another agency and, if so, we had to post the cheque to the invoicing company, but some invoices were paid directly without us having to forward a cheque.

At the time, I remember thinking about how anyone would manage this – or even figure out what they needed to do – if they were living alone, older, managing their care without assistance, did not speak English, or could not ask anyone. The process of receiving invoices, or cheques, and the payment methods were unclear to my family. There must be a better way of doing this. There was an urgent need for system design improvements.

4.3 ANALYSIS

In terms of outcomes, there are several of interest in this scenario. Typically, in an abstract case, we might look at patient safety, quality of care, or efficiency/productivity from the point of view of the healthcare operator. My natural interest in this case was skewed towards patient safety and overall experience, which for me included a sense of the psychosocial impacts of the various aspects of the scenario, and the extension of the impact on the family and carers of the patient.

There are three main themes that these experiences reveal and that reflect directly on human factors, ergonomics, and safety management. These are not complaints. I know that staff on wards are time poor, over-worked, and short staffed. This discussion is to raise issues for consideration when redesigning practices for continual improvement.

The key themes are as follows:

- Communication
- The safety reporting line
- Human factors in the systems of work

4.3.1 Communication

It's no secret that communication is important for safe healthcare systems (Iedema et al., 2019). I must say that we had some effective communication experiences in the intensive care unit and adapting some of those techniques in other stages of care would have been beneficial.

The following things might have helped on an extended hospital stay:

- *Patient-oriented notes:* A version of notes intended for consumption by the patient/advocate, or some process to ensure translation of notes at regular intervals. This would have meant that I would not have had to keep calling (and missing) the NUM for a complete file review every few days and we would have had verified information. It would have saved their frustration at me calling. Some healthcare providers are embracing this, in a movement known as 'OpenNotes',[3] and this has been shown to have positive outcomes for patients (Esch et al., 2016; Yu et al., 2017).
- *Improved written information in the patient's room:* There was a board in the room with patient goals written on it, but they were vague and distant (e.g., 'get out of hospital'). Having more real-time, accessible information about status, what to watch, what is a priority, what is uncertain, a collaborative digital 'scoreboard' of sorts (a term that would have appealed to dad's keen sporting interests) would have provided information, reduced anxiety, and helped progression. And it would have saved staff time, overall.
- *Communication/ward practices induction:* A basic tool for family and the patient that identified what the hospital expects from them when moving to a new ward would be useful. This would help so that we did not have to tiptoe around not knowing who to ask our questions. This could have clearly stated what equipment we needed to supply, who to ask for help, when they were available, etc. The healthcare educators and hosts inform students on placement. An extended reality and cloud-based accessible format could be effective so that family members could review this at home or on-site and become immersed in place and space.

There is quite a contrast between these kinds of proactive communication strategies and the one piece of information that we were consistently given: that we could call the hotline if we had concerns.

> There is guidance about communicating with patients (e.g., Australian Commission on Safety and Quality in Healthcare, 2016), which include recognition of language barriers and the importance of building trust, but what's missing is a sense of time/duration and context. The basic guidelines are great when thinking about communicating with one or two doctors about a diagnosis or procedure, but how does that work with multiple ward transfers, several shifts, and over an extended period, where trust may erode, and anxiety may increase? The guidelines do not inform the scenario in its wider context.

[3] www.opennotes.org/

4.3.2 Alarm Bells

I called the safety hotline and it caused welcome effects but the incident that prompted the call was not addressed in even a cursory manner. The hospital management teams committed to conduct a case review, but there was no follow-up on the incident. Dad was so concerned about where and how someone removed his lines. Ringing the safety alarm bell is not supposed to make people feel watched to fulfill the optics of a change in practice – it is meant to solve an issue and prompt ongoing, systematic improvements. I do not know if the administration recorded this incident and collected the hotline call data, or whether they reviewed the call as part of their quality management processes, but that is what should happen. It would be fascinating to review data on how often healthcare consumers ring such alarm bells, but I doubt that there are many organisations brave enough to share those data if they do collect and analyse them. All reports of this nature would be better shared, discussed, and used for learning and improvement, consistent with a basic risk management and safety management system (e.g., ISO, 2018, 2021). These issues may reflect problems either in design or scope of the reporting system, training and response strategy, or, of course, consumer misuse of the hotline.

Most organisations need to change how they perceive reports of concern – they are a gift: the chance of knowing a near miss rather than experiencing a loss.

4.3.3 Human Factors and Systems of Work

There were lots of complex systems and subsystems that we encountered through this experience. The hospital(s) exist in a wider health system and contain multiple complex subsystems (e.g., medical and nursing; administration, allied health, or services such as cleaning and food delivery). Models that we commonly use in human factors and ergonomics (HFE) such as the 'onion' model (Wilson & Sharples, 2015) are clearly relevant here and reflect the person-centred nature of HFE analysis. In such models, the person in question (usually a worker or system user) is at the centre of the system, linked to their equipment/technology, surrounded by their task and goals, their immediate organisational environment and work relationships, and the wider culture, practices, and management of their organisations, all of which is influenced by the external environment (including the economic, social, and political environment). Several versions of this have been adapted to put a patient, carer, or consumer at the centre of the system. The systems model used in healthcare human factors (the SEIF 3.0 model; Carayon et al., 2020) focuses on the journey of the patient and their caregivers over time and space, through the multiple systems with which they interact. This includes systems run by various organisations – for example, the hospital(s), the at-home rehabilitation programme, or the imaging company. Carayon et al. (2020) indicate how the patient experience comes from the combination and interaction of these various systems over time.

The duration of care is important to dad's scenario; it is not just about analysing potential errors in a 45-minute procedure, as we so often deal with in short case studies, but the cumulative, interacting outcomes, and dynamic inputs that affected his care and our experience. We often carve up our analyses of work systems in HF practice to levels of abstraction without venturing into the complex whole. I started to think about the HF issues for the patient in this experience, and the family/advocates, and seeing these roles, and their abilities and limitations as a particular instance of the system. For me, an important element of the SEIF model approach is the inclusion of families and friends in providing care: '*care is provided through a myriad of interactions between various individuals: the patients themselves, their families and friends, healthcare provides, and various other staff*' (Carayon & Wood, 2009, p. 29). They are part of the patient journey and the systems affect them. Designing healthcare systems that respond to the needs, abilities, and limitations of the patient and their family/advocates is a smart move, as family affects the patient's progress. And of course, good design of systems helps everyone, regardless of ability and limitation.

Some of the abilities and limitations for dad were related to his age, sensory perception (hearing), his acute condition (e.g., loss of muscle strength and mobility), and the outcomes of fatigue and distress. Also, he was in a state of anxiety and confused and disoriented by the space. Subsequently, he lacked trust in their systems and the hospital staff informed us about dad's progress in disjointed ways. Though these are outcomes of the system, they function as limitations for ongoing recovery tasks.

While there was attention to some human factors issues in dad's care, mostly related to mobility and the need to regain muscle strength, there was little focus on how factors affecting communication and information could affect his mood, attitude, expectations, goal-setting, and motivation. In other words, the administration ignored the system's design of care and its psychosocial impacts on my dad and on us. There was at times some acknowledgment of these issues (e.g., 'he's tired, that may be affecting him'), but no strategies to deal with system-based origins of some of these impacts.

> Psychosocial impacts of long-term hospital care are not something that the administration can simply solve by wheeling in a counsellor, despite these professionals being critical elements of healthcare systems, just as managers cannot simply solve psychosocial impacts at workplaces by referring people to employee assistance programmes (e.g., Caponecchia, 2019).

While we as family/advocates experienced some limitations (e.g., dexterity, overload, stress, and fatigue). Part of our role, and really the only thing that we could do, was to buffer against dad's limitations in the system. We could do this by being present, providing social support, supplying clothes, helping with meals and daily tasks, and importantly, sourcing and communicating complex information. In addition to designing for patient limitations, anything that helped the family/advocates to have an easier, more user-friendly, and less stressful experience in the system could help the patient in the system. Designing to make the systems easier to use, less stressful, and more user-friendly can impact everyone – and in this case, the system includes patients and their family/advocates because they have a key role in care delivery. This is a situation in which making the systems better for everyone can double the benefits for patients and make things easier on the staff; as such, the hospital can improve the direct and indirect healthcare experience for care recipients and their families. I hope that our experience can help identify ways to change the various systems with which we interacted to improve others' experience. Small, meaningful changes created by a systems model that considers the whole patient and family journey can make a significant impact.

4.4 REFLECTION AND CONCLUSION

After several long, stressful, worrying months, dad returned home and was able to access an excellent integrated care package for the following three months (with occupational therapy, physiotherapy, community nursing, etc.). He was extremely lucky to survive, adapt, and improve, with support from mum and the rest of our family. He uses a walker, has returned to driving, shopping, and daily walks, and completes all his own self-care tasks. We learned a lot from this experience. Complex systems are difficult to change, but this starts with trying to understand the systems and share experiences. This highlights the value of storytelling, which is hopefully something to which this reflection can contribute.

REFERENCES

Australian Commission on Safety and Quality in Healthcare. (2016). *Patient-clinician communication in hospitals: Communicating for safety at transitions in care*. Retrieved February 2, 2023, from www.safetyandquality.gov.au/sites/default/files/migrated/Information-sheet-for-healthcare-providers-Improving-patient-clinician-communication.pdf

Caponecchia, C. (2019). Risk management and bullying as a WHS hazard: Prevention of workplace bullying through work and organisational design. In P. D'Cruz, E. Noronha, C. Caponecchia, J. Escartin, D. Salin & M. Tuckey (Eds.), *Workplace bullying: Dignity and inclusion at work*. Springer.

Caponecchia, C., & Wyatt, A. (2011). *Preventing workplace bullying: An evidence-based guide for managers and employees*. Allen & Unwin.

Caponecchia, C., & Wyatt, A. (2021). Defining a safe system of work. *Safety and Health at Work, 12*(4), 421–423.

Carayon, P., & Wood, K. E. (2009). Patient safety; the role of human factors and systems engineering. In W. B. Rouse & D. A. Cotese (Eds.), *Engineering the system of healthcare delivery* (pp. 23–46). IOS Press.

Carayon, P., Wooldridge, A., Hoonakker, P., Hundt, A. S., & Kelly, M. M. (2020). SEIPS 3.0: Human-centred design of the patient journey for patient safety. *Applied Ergonomics, 84*.

Esch, T., Mejilla, R., Anselmo, M., Podtschaske, B., Delbanco, T., & Walker, J. (2016). Engaging patients through open notes: An evaluation using mixed methods. *BMJ Open, 6*.

Iedema, R., Greenhalgh, T., Russell, J., Alexander, J., Amer-Sharif, K., Gardner, P., Juniper, M., Lawton, R., Mahajan, R. P., McGuire, P., Roberts, C., Robson, W., Timmons, S., & Wilkinson, L. (2019). Spoken communication and patient safety: A new direction for healthcare communication policy research, education and practice. *BMJ Open Quality, 8*.

International Standards Organisation (ISO). (2018). *ISO31000 risk management – guidelines*. ISO.

International Standards Organisation (ISO). (2021). *ISO45001 occupational health and safety management systems: Requirements with guidance for use*. ISO.

New South Wales Clinical Excellence Commission. (n.d.). *REACH*. Retrieved February 2, 2023, from www.cec.health.nsw.gov.au/keep-patients-safe/reach

Tuckey, M., Zadow, A., Li, Y., & Caponecchia, C. (2019). Prevention of workplace bullying through work and organisational design. In D'Cruz, P., Noronha, E., Caponecchia, C., Escartin, J., Salin, D., & Tuckey, M. R. (Eds.), *Workplace bullying: Dignity and inclusion at work*. Springer.

Wilson, J. R., & Sharples, S. (2015). *Evaluation of human work* (4th ed.). CRC Press.

Yu, M. M., Weathers, A. L., Wu, A. D., & Evans, D. A. (2017). Sharing notes with patients: A review of current practice and considerations for neurologists. *Neurology Clinical Practice*, 7(3), 188.

5 Constructing Health by Gut Instinct

Back to Basics

Amy

This author dedicates this chapter to the child at the heart of this story, and the friends and family who cared – they offered and gave so much. Also, the gym friends, the swim coach, the friends-as-advocates, the friends-as-researchers, the many shoulders offered to lean on, and those who prayed for us. They helped us have faith in the universe once more.

My son Levi[1] was 12 at the time of this account. His primary gastrointestinal symptoms that inform this story began when he was 11. In fact, once I had researched past emails revealing my communication with family and his school, I found a trend of stomach-related issues starting from age 6, a few months after the sojourn of sole parenting began (Levi was 5 when his father left the household). The first time that he had stomach cramps was when Levi's father reneged on his promise to collect him and take him to the airport to meet his half-sister who was visiting for the holidays. Levi learned of the betrayal once his sister had already arrived in town, days later. He crumbled with the realisation and was at a loss to understand the broken promise. He had stomach cramps for ten days following. And so it goes: a gut–brain connection that creates mysterious illnesses or lowers the threshold of tolerance (Antonovsky, 1979) to give rise to an interruption caused by other nasties, like viruses or bacterial imbalances that thrive when the conditions are right.

Levi's timeline revealed intermittent stomach illnesses, but no more than a parent might expect when playing with other children attending a primary school and the related exposures to the nefarious mystery illnesses and infections that seem to cling to youth, albeit briefly. Levi had not established a recognisable pattern of ill health. However, by age 11, Levi's body started to reveal more that was worthy of study. He suffered from sinus infections and, by this time, increasingly erratic and episodic gastrointestinal distress. I remember that he started belching. Not burping, as a child might, but belching like a fire-breathing dragon. Not all the time but, when he did, it was severe. There were some weekends when he had diarrhoea, others when he vomited or felt nauseous. There were persistent, unexplained occurrences of generalised feelings of malaise or simply not feeling well, but there were gaps when he seemed

[1] A fictitious name has been used.

DOI: 10.1201/9781032711195-6

fine: many times, he was asymptomatic. The backdrop to this was the emergence of an athletic child, actively involved in sports like rugby, Surf Lifesaving nippers, swimming, karate, and little athletics. He loved his scooter play also. He ate well and had a moderate appetite which, sometimes, was robust. Fresh fruit and vegetables were part of his daily nutritional regimen. It was reasonable to believe, at least at first, that the episodes of illness were related to the various 'bugs' that children pick up in the school yard, but the persistence with prolonged and repeated experiences became concerning. He had been to our local general practitioner at least four times during the first year. Once, the doctor prescribed a broad-spectrum antibiotic. He had an ultrasound to check his appendix, and this was negative. Levi did breath tests to check for *Helicobacter pylori (H. Pylori)* (though the laboratory phlebotomists told us later that this test was contraindicated for a child at this age) and he was examined by ultrasound. Levi underwent general lab work for stool, urine, and blood samples (this was significant for a child of this age). During one visit, the doctor instructed me to teach Levi deep low belly breathing to promote relaxation and that was the extent of the doctor's advice during this consultation (given that I teach yoga, this was not too remarkable or new to me, just disappointing that this visit did not warrant the doctor's committed investigations of related biological causes). Levi missed days at school, a school camp, and he missed his favourite rugby sport practices and matches. The symptoms worsened. Was there exposure to household moulds? Could he have contracted a bug when travelling the year before in Laos, with river play and flying foxes? Could a poorly maintained water filter have exposed him to bacteria? Was he one of those children who was sensitive to food intolerance, or reactions to common bacteria that affect some people more than others?

Fast forward to three months before his 12th birthday, a September birthday child. By June, this child's body was screaming for the right attention. His symptoms worsened with a symphony of complaints, competing with one another simultaneously. He suffered belching, mouth ulcers, and vomited eight times per day. Levi endured stomach cramps that caused writhing pain. He was nauseous from his wake to his slumber; had water-like diarrhoea up to five times per day; had a bloated belly, especially after eating. He suffered leg muscle cramps with electrolyte imbalances, and he presented with random skin rashes. He experienced daily headaches, and, of course, was a casualty of ongoing malaise and fatigue. Levi consistently complained of right lower quadrant abdominal pain, though his entire abdomen caused him unendurable discomfort. We kept attending the doctor's office diligently, and requested a referral to a specialist paediatric gastroenterologist, a dietitian, and a psychologist. We attended eight general practitioner (GP) visits to this medical clinic during this time, with a phone consultation also. The primary GP and the practice medical director had seen Levi.[2] Levi undertook more lab work, at more expense to me. The lab tests were not definitive or conclusive, but they were not normal, with positive parasite of *Dientamoeba fragilis*, equivocal results for *H. pylori*, and elevated IgA and IgG, protective antibodies.

[2] Notably, the medical director saw mum in her yoga outfit and enquired in a presumptuous manner if the child was on any 'special diets' (assuming something like veganism was associated with yoga and, in turn, associated with the child's poor health presentation).

By August, Levi was in total bed rest for the month. When we saw the dietician she said that she did not understand why we were there to see her. We explained that with a long waiting list for the gastroenterologist, the general practitioner advised us to start with her as soon as we could. She did not believe that his symptoms were consistent with typical food sensitivities. We tried gluten-free, diary free, and low-FODMAPS diets.[3] I was ordering home-prepared special meals from a colleague (also trained as a dietician) who made meals that complied with these dietary constraints, and this helped. The psychologist followed. Levi also saw a naturopath and holistic chiropractor. At his worst, Levi could hardly lift his head from the bed or couch. I was scared. I wondered if my child would live. I started to imagine the worst. I prayed that Levi's experiences were a chapter, not a change in story.

When I called the GP's office, it took ten days (about 1 and a half weeks) to receive a return call. I called a friend who advised the 24-hour nursing triage hotline, and after the nursing assessment, she advised taking Levi to the emergency department of the local hospital. This would normally have made sense; except for a taxed hospital system mid-pandemic, and resources were scant. We tried the hospital emergency department, and this was a colossal failure. From intake to triage, admission and nursing staff told us that we were COVID-19 infection risks because Levi had a headache (as low hydration causes). As such, they asked us to wait outside on the footpath, sitting on garden chairs for hours.

The hospital could not advise when a doctor and the nurses would see Levi and the nurses could not determine the need for intravenous hydration therapy. I was concerned. The hospital wanted time to prepare a protective sterile room. I wondered how this was not already part of the routine preparations since the pandemic was late in its second year of effect. Their focus was on communicable disease, and not the symptoms that Levi presented with in relation to his gut health. Eventually, hospital staff moved us to a partial ward, partitioned from the rest of the hospital, replete with locked doors. The nursing teams locked us in with other families, none of whom were in protective personal equipment, but the staff stormed through in their complete protective gear. I felt vulnerable and my son was tired: if the staff genuinely believed that any of the families presented with infection risk, we would be in fertile environs for cross-infection, yet the staff protected themselves with personal equipment. I asked again about receiving attention for my son and a nurse told me that someone would address us soon. When 'soon' became another half an hour and there was no address, I tried to explore our options. However, the staff locked the doors, and they were amid shift handovers. I repeated my line of questioning with staff from the next shift and, again, they could not tell us when a doctor would see him. More than 4 hours later, I self-discharged my son, and we went home, exhausted.

It was a stressful experience, as compassionate as the nursing staff tried to be, a nurse scoffed at me, 'Well, you chose to come to the emergency department!' I replied that a triage nurse advised me to do so, and that this would not have been my preference. My son was very unwell, but we felt worse sitting in this ward, hours at

[3] FODMAPS are short-chain carbohydrates that may not be absorbed well in the gut, and they include foods that are fermentable, oligosaccharides, disaccharides, monosaccharides, and polyols. Refer to: www.monashfodmap.com/ibs-central/i-have-ibs/starting-the-low-fodmap-diet/

the hospital, with no timeline in sight for further skilled assessment or intervention. We left, and the experienced deterred my son from hospital admissions. Ironically, when I spoke with the specialist and the GP, they gave advice, 'go to the local hospital if you are concerned'. It was a ceaseless loop of poor options and no real solutions: a track that kept going around.

I brought Levi for an overnight stay in a larger city a couple of hours away from home and a paediatric gastroenterologist saw him in hospital the next day for advanced investigations. This was soon after the emergency department experience. The specialist advised him to have an endoscopy and colonoscopy. This was not an easy transition. He had to have a clean-out prep and he hated the experience – poor-tasting powders and quick runs to the loo with painful diarrhoea. He was in tears. I sat him on my lap, soothed him, and had to be 'cruel to be kind' to force the ongoing ingestion of the prescribed prep. Nervous, I knew that the investigations detected cancers and severe intestinal disease. My grandmother, Levi's great grandmother, had Crohn's disease. I hoped that this was not his experience, yet I also wanted to find something, to conclude the unknown, and to begin a constructive path forward. To my relief, the results showed no intestinal disease, but they were positive for stomach gastritis. The biopsies did not reveal any gluten intolerance and were negative for *H. pylori*. I was told by the paediatric gastroenterologist that he only had 'a mild condition'. When I returned to the GP, and my son was still symptomatic, albeit with modest improvement after the colonoscopy, this led to my dismay. He had started natural remedies, but the effects were modest. I wanted to know: What caused the gastritis and what could we do to help him return to the full-of-verve happy child that I knew? His gut had depleted his health.

Levi missed the entire term of school, and I helped homeschool my son. The school required daily notice of his homestay status, and this was yet another task on my 'to-do'. I walked the dog, attended several health-related appointments each week, and prepared special meals. The care was on my shoulders, solo. The father was absent during this time, including his time away for a five-week holiday. He stopped by for less than 20 minutes to see Levi during his recovery after his colonoscopy and headed on holiday that evening, calling Levi that night to chronicle the construction of his campfire. While we attended a naturopathy visit, the father's girlfriend sent me photos of their campsite while Levi was symptomatic. This was not helpful. During this time, I worked (hard), but had to defer most projects, which meant that business interruptions slowed my cashflow. The responsibility of sole parenting is rewarding, but also a summons of resilience (especially while supporting an unwell child).

I was trying to help a child manage chronic inflammation and congestion. I wanted to understand the symbolism at a psychic level and the science that contributed to these experiences. The medical team looked at me with pity, as though I were a mother who needed to cope better and accept the status of parenting a child with a disability. They suggested that the psychologist might investigate underlying issues and help with the transition to his acceptance of being unwell. They asked me, and I had already asked Levi, whether any adult had touched him inappropriately, in case abuse was causing anxiety and, in turn, triggering poor gut health. Thankfully, and he swore this to be the case, he had not. After three sessions focusing on mindfulness, boundary setting, and relaxation methods, and once learning of the organic causes of Levi's distress, the psychologist did not believe that he needed to continue with her sessions, but she invited us to resume services at any time that Levi should wish.

In the meantime, the medical staff appeared weary of hearing from me, and I would not let up my questions, research, and requests for investigations. I called, sent research papers, and emailed queries. Irritable bowel syndrome, the doctors told me, could create hyperactive motility of the gut, a gut–brain disturbance, with its own neural network. The doctors diagnosed Levi with visceral hyperalgesia. They had created a narrative to explain Levi's symptoms related to an extra-sensitivity to pain throughout his viscera (internal organs). The doctors referred Levi to the persistent pain clinic in the city, hours away, for outpatient appointments. I tossed, turned, and lost sleep. None of this felt right and it was not acceptable. I could not believe that my son's health had flipped so remarkably and his was a change in story. I spoke to family members (interstate and overseas) and several friends, all of whom I am grateful to for their listening ear. I received supportive calls and shared conversations with teachers, the vice principal, and the school chaplaincy.

His new school (high school the next year) arranged an appointment for me to discuss the matter with their student coordinator. To my relief, she was supportive (I feared that his admission, for which we also struggled to ensure within a competitive market and few places available, would be in jeopardy). Friends from the gym offered support, advice, and even places to stay if he needed to attend a city hospital again. The local pharmacists and their staff were supportive, with practical advice about what to take over the counter and what to expect, sharing personal stories too. Levi's swim coach reached out and offered for Levi to speak to him at any time, about any issue requiring the wisdom of a supportive man. He shared his story about his teenage health difficulties, and how he managed to rebuild his health and fitness to a competitive level with the help of his persistent and caring mum. This was inspiring and provided positive role modelling. I was grateful, as was Levi. Without these people, my isolation and the battle to construct my son's health while remaining in the dark would have been more difficult to bear.

When my son continued to suffer belching, bloating, nausea, vomiting, diarrhoea, mouth ulcers, cramps, and skin rashes, but the GP told me that his labs were not conclusive and definitive, and he 'looked fine' to her, I nearly collapsed. His right lower abdominal pain persisted, and I asked if she could palpate, feel anything, even though the ultrasound showed no appendicitis. This doctor told me that there was nothing wrong with my son. I stuttered and stammered:

> This is NOT my son. He can attend a ten-minute appointment here, but he cannot withstand much more. He will fall asleep in the car when we leave. He can barely walk the block without vomiting. He has not used his new trampoline. This is not the 12-year-old boy I know. What about his symptoms? What about my report; I live with him, and I have been caring for him?

The general practitioner told me that there was nothing else to do. I could manage the symptoms with over-the-counter pharmaceuticals and deal with his pain via psychology and the city-based pain clinic. If I could not manage, again, I was welcome to take him to the hospital or to the emergency room. The paediatric gastroenterologist specialist advised that another investigation of the transverse colon, via magnetic resonance enterography, could be undertaken, 'but I do not expect to find anything, and it is not medically indicated', she said. Why bother? The hospital experience and the trauma of preparation dissuaded my son. Swallowing dyes and undertaking more investigations were not attractive options when results did not substantiate their need: these circumstances stymied me. Notably, I asked about getting a second opinion from my friend's stepfather, another local GP and past medical educator, who I knew loved the academics underpinning medical diagnostics. I was told, 'oh, no, they're only going to repeat these tests, and this is unnecessary'.

I went home that night highly distressed. It took two days, and then it hit me – hard. My first experience with a monster migraine from the pit of Satan: blurred vision with fuzzy lines, a severe headache, nausea, and drunken fatigue. I could not drive, I could barely walk, and engaging with devices for more than 2 minutes made me collapse. My son and I shared the struggle of basic daily living activities, and I mostly slept in between. I cancelled my work, important work, that I would never have excused, but I could not help the circumstances. I was not safe to drive, and I could not engage. As I crawled my way back to the land of living and tested myself for COVID-19 antigens, just in case and because it was the right thing to do, I vowed to change healthcare practitioners. The statement 'He looks fine to me', compelled me. I needed to hold the diagnostic crystal to the sky and let the light refract differently, get new perspectives, and discover another way.

I declined the referral to the pain clinic. When I asked what they would do, I was told, 'Review his pain medications'. 'He is not on any', I said, 'and I do not want him on any'. I wrote to the GP's office and explained again why they were not on the right track. I was given a message to please make an appointment for 'a non-urgent appointment'. I was in no hurry to return. Instead, I booked appointments with an allergy doctor and a functional gut doctor (two different GPs at separate clinics). I was on the waiting list for another integrative health medical practitioner. The gut doctor I discovered through reverse engineering: I found a friend of a friend, interstate, who had experienced gut issues, received some remedies after testing at a special lab, not those routinely used by GPs. I contacted the lab to discover who from our area referred to this lab. I found this doctor on the list and made the appointment, though there were three weeks to wait. In the meantime, the holistic chiropractor had seen Levi. His heuristics instructed him to palpate Levi's stomach and he noted that his transverse colon seemed markedly distended and he thought that the family-provider team warranted further investigations. The allergy doctor confirmed that Levi was not well,[4] and his symptoms provided empirical evidence that providers should not ignore. She decided to try the antibiotic treatment immediately for *Dientamoeba fragilis*, given that some people are sensitive to this exposure. She prescribed treatments, also for his skin rashes and sinus (turbinates, which were still mildly troublesome). When we saw the functional gut doctor, he advised us to take an X-ray. Soon after, we learned that Levi's severe faecal impaction obstructed his colon from slow-transit constipation, and *not* a hyperactive gut that the previous doctors

[4] Relief, much that it was, to receive confirmation that we were *not* crazy.

told me that he had. The allergy doctor read the X-ray and explained this to us. I did not know about 'overflow diarrhoea', but I soon learned that this was possible: the gut attracts and expels water that pushes against an impaction, causes pain as it does so, and stool does not form because of the lack of space and time in the gut. Confoundedly, Levi had never had a history of traditional constipation, nor was he distracted by social play, as some children are, that he did not want to or feel the urge to go when needed. This was serious: an obstruction can cause a perforated bowel, severe dehydration with kidney failure, sepsis (a fatal blood infection), electrolyte imbalances, abscesses with further infection, and other such complications or causes of fatality (Smith et al., 2022).

I had taken to arranging alternating medical appointments, to provide a system of checks-and-balances and divergent thought that could be synthesised for logical patterns when synergistic thinking arose, or when it simply made sense to me. I assumed this management role because there was no overarching case management or adequate collaboration without my drive, though I wanted a carer to carry me, and I needed support. Where were the services like the Mayo Clinic in the United States, where doctors of complementary but distinct training evaluated a client and conferred to establish defensible decision-making while constructing a concerted action plan? I insisted on remaining a collaborative part of this new discovery team, and always tried to involve myself as such, even from the beginning when my ideas and reports were mostly discounted. The new health care team advised oral medications, ongoing naturopathic remedies, and tinctures. They helped a little. One time, Levi had an allergic reaction when the regime changed with a more rigorous clean-out effort. I deferred Levi's birthday celebration twice and eventually hosted this, even while he was only moderately well. His symptoms had only mildly improved. At one point, reverting to the past specialist's advice, we had a prescription of Endep to try (a medication given for depression, multiple sclerosis, and irritable bowel syndrome). 'I will accept the script, but please do not be offended if I do not fill it', I said, 'I want to see how Levi responds when we know that he is fully cleared of his impaction, and then reassess the need'. Thankfully, the script was never filled.

Our goal was to get him to school in term 4. We did not make it, but we did make graduation day. How? By taking things into our hands, which required a salutogenic approach to construct health (Mittelmark et al., 2017), rather than searching for severe disease. It was unacceptable to me for Levi not to be able to play, engage in sport, walk the dog, or attend school. It was unacceptable to me for him to suffer severe and chronic fatigue, nausea, and belching, or to be in tears with cramps, at times doubling over with the effort to constrain the discomfort. Inspired by conversations with the chiropractor and, importantly, through nightly guidance by an executive nursing operations manager friend of mine who was interstate, I sought help from a colonic clinic manager, bought an enema kit, and learned to administer home-based water-and-chamomile tea enemas. The clinic lady was sympathetic, and patient. Her advice was useful. The administration was clumsy at first, but the effect was near immediate. The next day, my son was releasing his impaction. Never had we celebrated poo the way that we did that day. The nightly conversation, 'Did you have your bowel movement today? How many times, dear?' was a near-joke in our household, yet it was also a serious investigation. Eradicating poison in the gut is a battle at a microscopic level of the microbiomes – changing the entire ecosystem internally and riding into battle to do so, accepting nothing other than victory. The improvement was gradual, but not inconsequential. I asked the GPs about booking a colonic and was told, with caution, 'I would not do this, it should be medically supervised, and nurse administered'. 'Okay, where, when, and with whom?' I asked, 'Could this be done in the hospital with a strategic admission, or via home care nursing?' However, GPs cannot direct the hospital admission. The specialist previously referred me back to the GP when I had asked this question for Levi's triage and care, and nobody had any case management directive as to how to seek this medical care. I persevered and after intermittent home enemas, we started to do these nightly. Virtual nursing care by phone from my friend, who continued to provide instructions while I administered the treatment, the company of a dog who needed to be part of the party, a curious cat who wanted to play with the lines, and a brave boy who knew that as humbling and challenging as this was, he needed a new path towards health. Importantly, a follow-up X-ray showed that two-thirds of his impacted bowel had completely cleared (yay us!). However, the ascending colon, notably the right lower abdominal quadrant that was the repeated cause for concern by my son when we had seen the doctors previously, was still significantly compacted.

A phone consultation with the functional gut doctor resulted in a revert of the advice. Levi was improving but the dragon continued to roar, and he was still belching. The GP advice that evening was, 'Book the colonic'. 'What?', I was confused. I had presented that in my line of questioning during the last consultation with the GP and the same doctor told me to refrain from this, causing me to cancel Levi's colonic appointment. 'Go for the colonic, it will be okay,' he said. And so, we did. More tears, more care from mum because the practitioner does not insert any tubes, the family member does this for a child (that would be me again). An hour session, more time on the loo before we left, and a night of exhaustion. The next morning was a miracle. 'Mum, I feel NORMAL!' a smiling boy exclaimed. The look of joy on Levi's face in that moment is one that I will relive. Levi hugged me and held me strongly. He jumped out of bed, skipped down the hallway, giggled, and hugged me again. I was in disbelief. He had no nausea, no discomfort. He felt light and great. I was in a daze. Had we really gotten there? I took the day off scheduled work, and Levi and I went to the beach. We had fun. He played. I called several friends. I cried, this time with happiness, through my exhaustion. Five intense months, longer still per the path of incorrect medical diagnostics, significant expenses managed solo, the isolation, the angst, and the struggle: I felt it all move through me in waves.

Success is sweeter when borne from struggle. Through the generosity of friendships and listening ears, my life was made more buoyant when I needed it most. That night, Levi made it to his graduation ceremony. He had not learned the class dance or the stage routines, but he made it there. He has been well since, although it was a slow reconstruction to return to his full fitness and stamina. We continued the natural remedies that the functional medicine practitioners prescribed and reduced the stool-softening laxatives to an exceptionally low dose, with less frequency, gradually eliminating use. Levi needed another four months of gut retraining for the natural peristalsis (gut movement) and the rebalancing of his microbiome (healthy gut bacteria).

Notably, the clean out preparation of the colonoscopy did not make enough difference – he improved afterwards, but the doctors were not treating the poison, the gut ecosystem, or the allostatic, regulatory functions. Levi's gut distention and hypo-motility were factors that those doctors had not identified, so there were no intervening strategies, and he suffered gut impaction again. However, with the latest remedies, Levi was on the mend. He consistently ate well, and he maintained his hydration. He retrained for swimming to attend Nipper's surf sport for less than half of the remaining season (in fact, only a couple of sessions remained). However, these were good problems to have: how to get fit during the summer holidays with the return of his health.

I am grateful. Near the end of our health discoveries, Levi attended an integrative, functional medicine health specialist, who had mixed training and education as a nutritionist and a genetic health researcher. This provided us with more reflective analysis of life events and health markers. Also, the functional medicine nutritionist prescribed remedies to assist with the reconstruction of Levi's complex microbiome health. We continued the dietary review to help better understand the interacting factors and his threshold tolerances to nutrition and stress events. We did not want a repeat of this occurrence. Levi maintained his relaxation strategies, like baths with Epsom salts and aromatherapy, delta-wave music, homeopathic remedies, foot and hand massages from mum (and sometimes Thai massages, when he was not too giggly), nature hikes, and deep diaphragmatic breathing. I am pleased that this young man is learning how to set boundaries, foster effective social relations, and discard circumstances that no longer serve him. Through this experience, he understood the benefits of restoring his well-being through mindfulness practices in a profound way and he learned more about the benefits of sound nutrition.

I wrote to the doctors at the original medical clinic and to the specialist and told them of our findings. I shared the 'before and after' photos (they were remarkable). I had previously recorded my son's wails at night, to depict the severity of his symptoms. I explained that I learned that the X-ray should have been part of the first-line investigations. Correct diagnostics and early intervention could have saved us from significant angst and expense. It would have improved Levi's school and sport attendance. Had the doctors ordered the right investigations and helped us better manage his symptoms, Levi would have been well sooner. Also, the medical approach could have been improved by looking at what constructed his health. Instead, they examined his symptom presentation through a fractured lens of disease states which led the practitioners down a rabbit warren of misdiagnosis influenced by their cognitive biases (Besnard & Robson, 2010). I gave my thanks for their attention but added my hopes that this information could help their diagnostics if they come across this again among other children. I did not receive a response.

Through this sojourn (and we are with blessed gratitude that this was a chapter in life, not a change in story), we learned:

- The traditional medical system is in a shambles. The corporatisation of healthcare with considerable time constraints coupled with practitioner egos and fears about the medicolegal implications of their care dim their creativity. These factors sway cognitive biases. They detract from the combined scientific and intuitive healthcare approaches that can inform sound diagnostics.

- Medical practice is just that – practice. The consumer as 'customer' pays the practitioner handsomely for this practice, no matter what. There is an expected community standard that a healthcare provider is beholden to the delivery of quality care, yet there is no service contract against which to hold them to account.
- Remember that you can improve health practices, so challenge what does not feel right: doctors are not Gods, and providers must adopt a participatory approach to involve the care recipient in the delivery of their care.
- If lab work is not definitive or conclusive (and yet not normal either), go back to the empirical evidence of client symptoms. Rely on intuition and the discipline of deductive reasoning to rediscover the art and science of medical care.
- Functional gut health is important; it affects mood, performance, energy, hydration, nutrition, soft tissues, and skin. If medical providers do not understand the nuances of constructing gut health, then they do not comprehend health.
- A pathogenic approach, investigating disease states, versus salutogenic approaches to inform the construction of health can cause passivity among care recipients who simply survive their lives, versus actively engaging in a life well lived.
- With anything medically complex, get ready to be your own healthcare case manager. Read, research, and keep investigating material to determine sources of truth. One must be prepared to partner with the medical team, though the team might not trust this approach, and, at times, they may shun your involvement. Keep finding the right people that welcome your contributions.
- Second and third opinions are useful.
- Schedule two providers if visiting doctors – alternate appointments for verification, validity, and action planning – one orders a test, the next one can interpret, they might start a plan, the other can refine it. It is all part of iterative (back and forth) design to get it right. Avoid service delays between appointments by doing this.
- There are waiting lists for medical care, long ones. Set appointments in advance and cancel them if you need to.
- Believe in yourself and keep questioning the advice if it does not sit well with you.
- Natural, old-school remedies can be useful – we learned to go back to basics: hydration, nutrition, sunlight, exercise, and 'clean the tubes' from top to bottom and back again, from both ends, when gut issues are troubling.
- Friendships matter: learn from your friends and their experiences and seek their support.
- Multimodal approaches can be effective: cast the net wide when you are not getting the answers that you need and learn from everyone. Discard what you do not need but take what you can if it helps the puzzle pieces fit just-right again. Humble yourself to tell your story if it means that someone will share theirs and you become better informed, and thus more empowered.

5.1 A YEAR LATER

Levi and I feel blessed by nature and the Gods (or whatever the beliefs) because of Levi's good health. Levi has continued his trajectory of well-being. He completed his first year of high school, during which he was an avid rugby team player. He no longer suffers symptoms of gut distress. He has gained healthy weight. He resumed swim lessons and qualified as a Surf Rescuer to help other children learn surf safety in their Nippers programmes. He has started wakeboarding. He uses the home garage gym for his daily workouts. When presented with disappointments from his father (who has been absent in his life for the last several years) or others, he is quick to respond: 'I am okay. I not responsible for their poor behaviours'. And he knows that his ill-health experiences led to this story that may help others.

REFERENCES

Antonovsky, A. (1979). *Health, stress and coping*. Jossey-Bass.

Besnard, D., & Robson, R. J. (2010). *Overlooking causes in healthcare accident analysis: Choosing the analysis is choosing the results* [Research Report CRC_WP_2010_1, hal-00505546v2]. MINES Tech.

Mittelmark, M. B., Sagy, S., Eriksson, M., Bauer, G. F., Pelikan, J. M., Lindstrom, B., & Espenes, G. A. (Eds.). (2017). *The handbook of salutogenesis*. Springer.

Smith, D. A., Kashyap, S., & Nehring, S. M. (2022, January). *Bowel obstruction*. StatPearls Publishing. Retrieved August 1, 2022, from www.ncbi.nlm.nih.gov/books/NBK441975/

6 Reflections on 'Gut Instinct' by a Work Design Strategist

Sara Pazell

I reflect on this story from the vantage point of work design, and I accept that it can be difficult to make sense of the sinuous tales of healthcare woes. However, a diagnostic pathway is not as straightforward as it may seem. People and their circumstances are complex. The quality of care relationships were diminished because of time pressures, the hierarchical power frameworks within medicine, and the stressors that may erode generosity and benevolence among providers (Edú-Valsania et al., 2022; López-Núñez et al., 2020).

I created a journey map (Rosenbaum et al., 2017) to depict the biased pathway of more than 18 months of Amy's family duress, poor diagnostics, and mistaken treatment planning (Figure 6.1). The involvement of a paediatric gastroenterologist in the trajectory of wrong diagnostics and treatment planning was concerning. In response, the general practitioner deferred to the specialist's advice and a pattern of confirmation bias was established (Saposnik et al., 2016), clouding effective care and management. The mum could not understand why the doctors dismissed her, and it took incredible strength during vulnerable and destabilising times to speak against their recommendations (like to attend city-based outpatient pain clinics that were hours away from home, or to undergo more invasive investigations that were doubted even by the specialist who made the recommendation). The blinders of partiality and judgement could have resulted in devastating effects with the death of a child. Yet the repair pathway was quick once the parent discarded the conventional medical advice. This does not mean to suggest that conventional pathways are wrong, but they were faulty in this case and the doctors failed to seek findings from an important diagnostic tool, an X-Ray, early in the investigations. Also, there were limitations in conventional remedies, because they failed to accept the simple healing power of old-school colonic irrigation. In this case, the parent deemed the benefits to outweigh the risks of home treatments (enemas) and clinic treatments (a colonic) with administration by the mum because the doctors offered no other solutions of merit.

 DOI: 10.1201/9781032711195-7

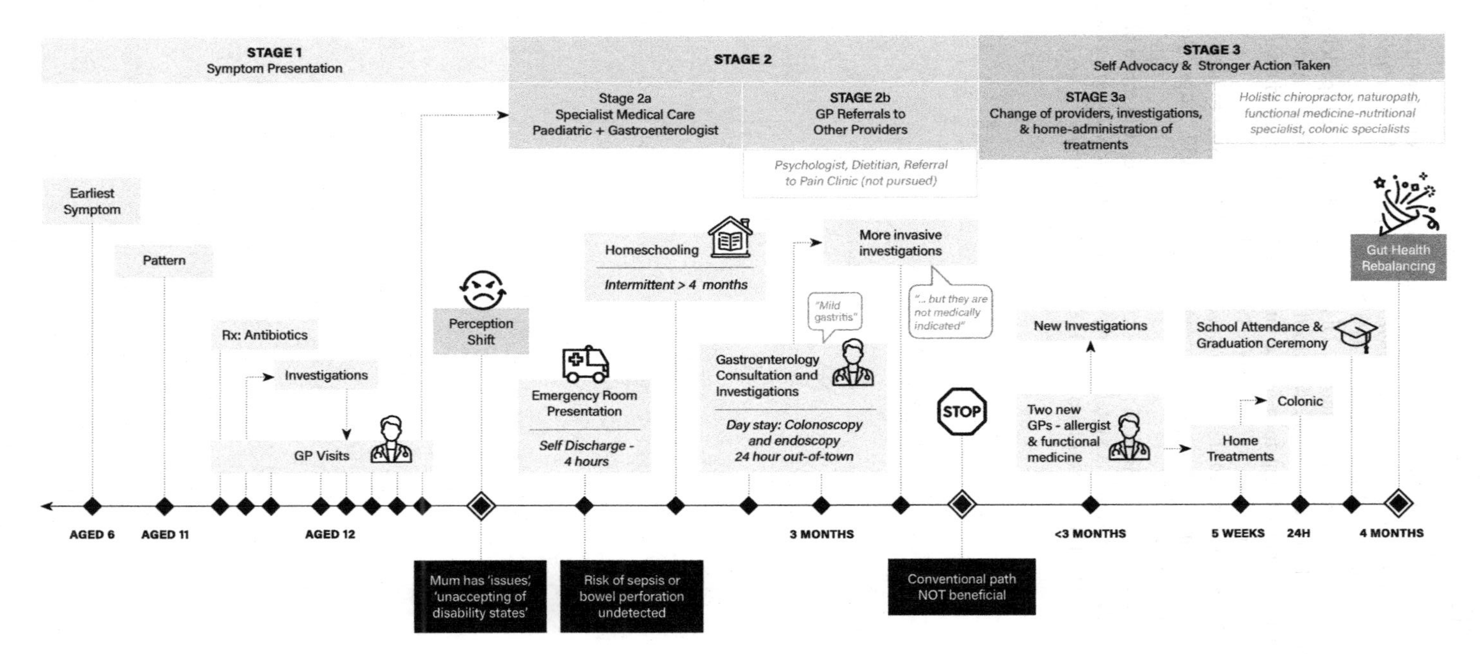

FIGURE 6.1 An 18-month health journey of a child with gut issues.

REFERENCES

Edú-Valsania, S., Laguía, A., & Moriano, J. A. (2022). Burnout: A review of theory and measurement. *International Journal of Environmental Research and Public Health, 19*(3), 1780. https://doi.org/10.3390/ijerph19031780

López-Núñez, M. I., Rubio-Valdehita, S., Diaz-Ramiro, E. M., & Aparicio-García, M. E. (2020). Psychological capital, workload, and burnout: What's new? The impact of personal accomplishment to promote sustainable working conditions. *Sustainability, 12*(19), 8124. https://doi.org/10.3390/su12198124

Rosenbaum, M. S., Otalora, M. L., & Ramírez, G. C. (2017). How to create a realistic customer journey map. *Business Horizons, 60*(1), 143–150. https://doi.org/10.1016/j.bushor.2016.09.010

Saposnik, G., Redelmeier, D., Ruff, C. C., & Tobler, P. N. (2016). Cognitive biases associated with medical decisions: A systematic review. *BMC Medical Informatics and Decision Making, 16*(1), 138. https://doi.org/10.1186/s12911-016-0377-1

7 Letters of Concern about Elder Care in a Hospital Environment

Miranda

The author dedicates this chapter to the parent (all parents) and the idea of ageism in healthcare delivery when risk-averse doctors and biased care teams make decisions that could derail rehabilitation of an ageing parent. During a period of hospital care in which he woke from a coma, my father (the ageing parent), told me of his dream state: he was walking around a tall building, there were no doors or windows, yet he heard us talking inside and he could feel our movement. He knocked on all the walls, but he could not get in. He desperately wanted to join his family.

7.1 EMAIL EXCERPT: LETTER 1 – SOON AFTER AN ACUTE CARE HOSPITAL STAY

Dear Hospital Administrator,

My father attended your hospital in the intensive care unit, the cardiac ward, and now he is in the renal care ward. He was in hospital because of cascading medical events that included his recent history of a) anaemia, b) a staph infection (contracted while in the hospital previously), c) blood transfusions, d) a peripherally-inserted central catheter (PICC) line that became infected, e) contraction of Influenza A, f) viral attack of his heart, kidneys, and lungs that induced a heart attack, and g) pneumonia. Treatment during this hospital stay included his induced coma and paralysis with complex intervention to stabilise his vitals (dialysis, tube feeding, catheter assistance, etc.). Dad is doing better each week and is progressing daily. He is mostly lucid but has episodic delirium that could come from many sources (e.g., fatigue, medication-related, infection-based, or post-coma effects).

My mother appreciated her daily interactions with the care team because she found them to be friendly to her, thank you. However, we have had some concerns about the overarching care planning strategies. For example:

1) The doctors and care team considered my father as a candidate for referral to skilled nursing long-term care, per one of the attending nurses. I asked

DOI: 10.1201/9781032711195-8

about whether they had considered rehabilitation. Although his health needs were multi-faceted and complex, his significant functional deterioration is recent and was rapid. Before these events, he was independent at home with all daily living activities, except donning laced shoes over his left ankle-foot orthoses. He drove, attended social bridge events, and attended university work on a casual basis. Cognitively, he was performing at an elevated level.

2) After challenging the long-term care considerations of the care team, the medical care team referred my father to rehabilitation. A medical intake team refused his access to a reputable and preferred rehabilitation hospital. The social worker and care team told us that this was because he was unlikely suited to fast-track rehabilitation in the private pay system. The doctor subsequently referred him to the community rehabilitation hospital, hours away from where my parents live, and that would require at least three hours of driving a day for my mother. My mother knew nothing about this, and I asked whether the social work and care team consulted her on these plans. They had not. I asked the nursing team whether their care planning would usually involve the family members in these discussions. My mother lives in the Northeast of town and I requested a referral instead to an alternative local community rehabilitation hospital to support her access. I heard that it was a more modern facility with a better reputation for care than the other rehabilitation hospital. Initially, the social worker explained that the doctor had already referred my father to the other rehabilitation hospital. I questioned whether the referring doctor considered the impact on my mother, and I asked if the doctor could resubmit a referral request to the hospital that was closer to her home. Only when I insisted did the care team investigate these requests. As such, the doctor made a new referral to ensure that my father would be closer to their home.
3) I asked for a social work referral to help address my mother's questions about service and funding support. The nursing team did not realise that a social worker had not yet seen my mother, nor had the doctor made the referral.
4) My father was not eating much while in hospital. I asked the nursing team whether a supervised meal intake program could occur and if the dietitian realised that my father's nutritional intake was poor. I explained to them that he required skilled behavioural management with intervention to encourage his eating and that my mum should not have been left to manage this given the dynamics in their personal relationship (he got grumpy with her when she prompted his need to eat yet responded better to the instructions of nursing or care staff). The charge nurse told me that, yes, if my mother was in the room then the staff assumed that my father was fine with my mother's support. That was not the case, and he required skilled and supportive assistance.
5) I was concerned about the care team's reluctance to initiate a bowel management/continence program for my father. A few days ago, in the afternoon, I asked whether my father could remain seated in his chair, out of bed, to

receive my visit since I had just arrived, and he was still alert. Unfortunately, he was incontinent with faecal matter in his pad when I arrived. He wanted to try the commode chair for the toilet. The nursing team explained that they needed to bed-roll change him and that, if they were making the effort to put him in bed, they would not get him up again. As such, they did just that: they cleaned him and moved him back into his bed.

The next day, my father demonstrated his capacity to sit upright out of bed in his chair for at least three hours in the morning and then, after a mid-afternoon nap, sat up again on the edge of his bed for at least an hour. Once again, he asked to use the commode. I requested this support, but the nursing team refused this request. A nurse told me that since he was not aware of his bowel needs, they would wait until he messed his pads and clean this up using a bedroll technique. I suggested that his skin integrity would be better if he did not soil his pads. I argued that he would have a better chance to reset his neurological capacity for continence with habitual patterns of trying to use the commode half an hour after a meal. I explained, also, that a rehabilitation unit would receive him more favourably if he were continent. The nursing staff did not agree with this approach. I have asked my mother to check with the doctor the next day about whether the doctor could update his care plan to include bowel management. He was beginning to stand with the physiotherapist, after all.

6) The nursing team and physiotherapist explained that my father's care plan orders outlined that only the physiotherapist could trial the sit-stand and front-wheeled walker support with him, not the nursing team. However, the physiotherapist informed me that she had him push off the bed edge, reach the walker, and stand with contact-guard and stand-by assistance only, including verbal cueing to activate an extensor pattern to stand tall and engage his gluteal muscles. Surely the nursing team could have included this strategy for more repeat trials during the day and this may have aided his progression to transfer out of bed and, thus, teach him toileting patterns again?

These were the concerns that we had while he received acute care. I understand and appreciate the need to achieve medical stability, and I am thankful that his care will continue in a rehabilitation setting. I hope that this information will aid your quality review of care delivery.

–

This email was unanswered by the hospital facility.

7.2 EMAIL EXCERPT: LETTER 2 – WEEKS LATER, THE DAY OF DISCHARGE FROM REHABILITATION SERVICES

Dear Executive Director of State Health Services

My father recently experienced some complex medical events including a viral-induced heart attack, coma, and renal problems. The doctor discharged him today from the community inpatient rehabilitation hospital. We are thankful for the day-to-day care provided. However, I am disappointed by the poor communication and care planning that occurred in preparation for his discharge home.

My mother found out today, via a phone call with my father, that his physician told him that he 'was cleared to go home today!' The care teams held a case conference in the morning but without involvement of, consideration of, or notification to his family, including my mother who the medical teams expect will receive my father as his primary carer. In so doing, the care team set up expectations for my father that were impossible to re-set. He was excited to go home. There was no consideration of the steps that service agencies needed to qualify my father for transitional care and subsequent home rehabilitation care, despite my repeated requests. I reiterated these requests during his prior acute hospital ward stay. My father could have received social support to make an informed decision about his discharge planning that would have given him the best fighting chance for maximum rehabilitation. There was no consideration of my mother's capabilities to provide support to my father when he returned home in a deconditioned state contrasted with his previous independent living status (he was driving, providing research and teaching support at a local University, attending social bridge club events several days per week, and attending to his own errands. He was independent in his activities of daily living and instrumental activities of daily living). Instead, the doctor discharged him home without a preliminary home visit to determine his safety and performance contextualised to the home environment. In fact, I was told that the rehabilitation team's 'clinical reasoning did not determine the necessity for this to occur', but that they were sending him home with equipment! Certainly, a home visit could be useful to determine whether the equipment is suitable, a good fit, accessible for the user in their environment, in context to his performance expectations (For example, he typically showers two to three times per day, yet they have advised a home care assistant to help him transition to home with showering assistance up to three times per week). He is at risk for falls also.[1]

The referring physician only considered rehabilitation as a service of outpatient care (rather than the preferred home care option) which now means that my mother will be responsible for driving him to and from outpatient care three times per week, at which time he will receive two hours of care while she waits. Home-based rehabilitation would have been helpful, but they did not coordinate the paperwork and funding support required by the government agencies in a timely manner.

An occupational therapist told me that the doctor referred my father to a Commonwealth Home Support Program after I expressed concerns. However, when I asked the doctor about the service support or financial packages

[1] Postscript: As it turned out, he returned home without the shower chair that they said that he needed and, when delivered later, it did not fit in the cubicle. They also provided a front-wheeled walker, without knowing how many floor rugs were at home, or the width of the doorways, which led to difficulties in home access.

considered in care planning, he could not tell me. I had hoped that the discharging physician would be aware of the care planning to which he was providing consent. The occupational therapist, a member of the rehabilitation team, explained these last-minute service considerations, and the doctor announced the discharge orders on the same day. As it stands, we do not understand his eligibility or the scope of service benefits available through these programs.

When my father was in hospital, twice the doctors told me to expect my father to die from his health ordeals. He did not. He is here with us a little longer on this earth. I am thankful for the medical care. However, at the hospital, it seemed like the focus was purely on medical stabilisation, without focus on rehabilitation to help him achieve his premorbid functional capability. I had hoped that a rehabilitation hospital would be more effective in their consideration of capability and care planning, but it seems that a rushed approach to achieve 'near enough' was 'good enough' in this case. While the team tried to advance the position that a patient-centred approach considered my father's wishes to be home, they also needed to consider ways to involve my mother and my father in care planning, helping them to understand the service packages and eligibility criteria, and what would be required to optimise his access to the best possible progressive model of rehabilitation. Even the healthcare professionals exhibited their ageism, and they are the ones to determine that care dependence is a norm, rather than an exception. Also, they assumed that my mother would provide primary care for my father, not understanding her limitations.

I am concerned about my father's level of insight about his care needs, the demands that he will now place on my mother for her continual 24/7 support, his needs to be as fit as possible to achieve independence, and the risk of falls and ongoing debility and frailty that can arise without a proper care approach.

Thank you for considering my feedback.

The hospital administration provided me with a brief and general response to my letter. They confirmed their receipt of the email and stated that the staff believed that they always prioritised patient care needs. However, during the phone call about the discharge plans, the occupational therapist admitted that they could have better researched his home needs and the service support packages before suggesting this discharge.

7.2.1 Reflections

- Healthcare delivery seems constrained by time and rushed decision-making. The impacts of these phenomena are significant to the care recipients and their families.
- A person-centred, integrated approach to care would better support healthcare consumers and their families and ease the planning among practitioners. Walking and mobility should not be a task left to rehabilitation teams to facilitate for care recipients. These are not functions 'owned' by the boundaries of professional practice, but by a health consumer and their capabilities.

- When an older person presents with complex care needs, there is a risk that ageism will lend to the decision-making bias among the care teams. Too often, care providers assume that an older adult must be ready to transition out of home and into high-care facilities, versus participating in rehabilitation after an acute and adverse health event.
- The environment and its design impacts on occupant health. Healthcare intervention strategies must be contextualised to the environment: social and physical.
- Most families do not know what to ask, even when they participate in care planning. They need proper education to make informed decisions. The experience can be overwhelming to those who are unfamiliar with the complexities of healthcare management.

Part II

Practitioner Stories

8 A 50-Year Journey to a Concept of Health

David Beaumont

At the age of 11, I knew that I was going to be a doctor. I just *knew*. It's described as a 'calling' and many people in the caring professions have experienced that call.

Fifty years later, I reflect here on how my understanding of health and how my notion of how to be a doctor has evolved. It continues to do so. I see now that it has led me to a concept of health that is vastly different to that of my training. Even my *definition* of health has changed and, with it, the way that I see the relationship between doctors and patients. No longer do I refer to them as patients. Instead, I call them 'clients'. The word reflects my role in serving them, hopefully adding to their lives.

But that is true of this most recent phase of my career, which has evolved in stages. Before that time, they were patients, patiently in receipt of my ministrations. I was doctor, they were patient.

Evolution implies an increase in complexity. Expansion. Maturation. That is how I see what has happened in my relationship with my patients. In every phase, the relationship between us has evolved. So has the definition or concept of health. Inevitably, that also means that the outcomes have evolved too. This opens a whole new world of possibility. What potential does this create? Where could the practice of medicine go? What is the future role of doctors?

8.1 1979–1989: DEPENDENCE

The year 1979 saw me enter medical school at a university in the north of England. Oh, the buzz of excitement we shared on that first day. This was the first step to becoming a doctor. It felt as though we'd already made it.

School hadn't been hard. I had a job to do: study the right subjects and achieve as high a grade as possible. I was studying the sciences, which I loved. My sole focus was getting into medical school. Focus. Application. Get straight As. Job done.

The same applied during the first two years of medical school: anatomy, physiology, and biochemistry. I excelled. To me, medicine was a science. This was the heyday of medical *science*. Infectious diseases had been tamed and the epidemic of chronic diseases hadn't yet started. Cures for cancer and heart disease were surely just around the corner. It was only a matter of time.

But it was an illusion. It was the story we were fed; the expectation that we were given. It was wrong.

The reality of our position as medical students was made evident in ward work. There was a strict hierarchy. Medical students were the bottom feeders in the pond

DOI: 10.1201/9781032711195-10

of the hospital. Near the top of the food chain were the consultants; higher still swam the great white sharks, the professors.

But there was a form of life below us: the patients.

> This is a teaching hospital. It is a privilege for patients to be treated here. In return, they accept that they will be examined by multiple medical students. There's a pneumonia in bed 5. Go and listen to the breath sounds.

Patients were often not referred to by name. There was a certain convenience in being able to identify them by their diagnosis.

The power imbalance was evident. Doctors had the power; patients subjected themselves to examination, investigation, and treatment. The payoff was management of their condition, either to cure or control their symptoms. It worked – but it was paternalistic. There was a tacit acceptance of this by both parties.

The goal was health. And the definition of health? Health is *the absence of disease*. That was the result both parties sought. It was the doctor's responsibility to bring it about. Lifestyle factors were barely recognised. Even when they were, it was the doctor who decided. 'I'm not going to operate unless you stop smoking'.

But soon afterwards there was a turning point in health, disease, and the role of doctors. The purely medical model in which we worked was being challenged. Over the years that followed my medical training, it became clear that we were losing the battle against chronic disease. There was a growing epidemic of obesity, type 2 diabetes, and heart disease. New drugs were controlling them to a degree, but was this truly achieving the definition of health as the absence of disease?

Patients were dependent on doctors for the management of their new diseases. Doctors took responsibility for outcomes. In return, we received prestige, acknowledgement of our expertise, and financial reward. These were extrinsic motivators. The payback for our part of the deal was external validation. We were valued for doing a good (or good enough) job. Patients ceded control of their health to doctors, and just sat back.

I went straight from my junior hospital jobs to general practice training, as did many of my colleagues. I secured a partnership in a practice in a rural village in my home county of Yorkshire. At the time, I was thrilled to be selected from many applicants. It is only as I look back that I realise what a privilege it was, those 12 years that I worked there.

At the age of 28, I was a partner in a five-doctor practice. It was the only practice in a semi-rural village. This could have been the highest position I ever reached, the pinnacle of my medical career. Patients always called me 'Dr Beaumont'. There was profound respect for the doctors in the practice. We did our own out-of-hours emergency cover, 24/7/365. We never engaged locum doctors. Holidays were strictly rostered to avoid it.

That first summer, I was asked to formally open the village fête. In a sudden twist, a well-known Yorkshire comedian who had been invited before me but declined suddenly became available again. Did I mind sharing the honour of the grand opening with Charlie Williams? He was regularly on television at that time. He was a celebrity. But so was I, simply by virtue of my position.

My wife and I were newly married. We lived in the village alongside my patients. One evening there was a knock at the door. A woman carrying a small child was standing on the doorstep. Joanna had cut her foot; would I look at it? Nothing that an Elastoplast could not sort out. Although there was a sense of intrusion into our lives, we knew that it came with the territory. This was the role I and my wife had taken on.

On one occasion, I was cornered by a patient in the aisle of the only supermarket in the village. Would I give him a repeat medical certificate to save making an appointment? I would not, thinking it was one step too far, even in a dependent relationship.

There were rewards, such as when a patient came into my consultation room carrying what was obviously an expensive, boxed bottle of spirits. 'The dermatologist said I should buy you the best bottle of brandy I can afford. You've just saved my life'.

Three weeks earlier the man had come to see me with a cough and cold. Nothing of any significance.

As I listened to his chest I asked him, 'How long have you had this mole on your shoulder?'

'What mole?' he replied.

I really did not like the look of it, so I made an urgent referral. My suspicion was correct. It was a melanoma, a malignant skin cancer. It was caught early and removed, effecting cure. It gave me a warm glow. The gesture, not the brandy (though, that too). Had I done a good job? Of course, I had, and I could feel justifiably pleased with myself. But I was only doing the job I had been trained to do. My internal validation should have been sufficient.

That was one of many presents I received from my patients. At Christmas, every Christmas, one of the local farmers delivered to the houses of each of the five partners and their families a wooden box full of his produce. Vegetables so fresh the soil was still clinging to them. It was a wonderful expression of his gratitude for the care he and his family received from his family doctors.

What I did not see at the time was the fishhook hidden in this kind of relationship. There were rewards for a job well done, and gifts in anticipation of the same. But there was an underlying expectation that the arrangement *meant* that there would be a good outcome. My health is taken care of because my doctor is taking responsibility for it. And we accepted that.

But what if it went wrong? What if there is not a good outcome (because often there is not). God forbid, what if I make a mistake?

It was 6 p.m. on a Friday evening. I had just got home. But a patient was playing on my mind. My last patient of the day. She was a young woman with a headache; an 'extra', slotted in at the end of my session. I knew that I had rushed the consultation because I had wanted to get home to my family. I had two options; I could spend the evening wondering if there was something more serious going on, or I could check on her. I did the latter. I turned round, went back to the surgery, pulled out her (manual) records and phoned her. After two paracetamol tablets, she was feeling a lot better. She was clearly surprised to get the call. But at least *I* could relax.

The most memorable example of this phase occurred one Saturday morning. I was on call and was asked to visit a six-month-old baby with diarrhoea. The proportion of home visits to surgery visits was much higher in those days than it is now, and mother and baby lived just round the corner from me.

Baby Sophie was fine. She had a bit of diarrhoea, her tummy was soft, and she had just a very slight temperature. I prescribed some rehydration salts and advised mum to stop feeding her for 24 hours. Job done. I knew Mary well (as I did many of our patients). She was a sensible mother, and she was happy with the plan.

At 4 p.m., I received a call. It was Mary, the baby's mother. Sophie was fitting. I was round at the house in 2 minutes. The fitting had already stopped, but this had been no febrile convulsion. The baby was quiet, grey, floppy, her eyes sunken and scared. I can see her even now.

The ambulance was there within 10 minutes. In hospital, meningococcal meningitis was diagnosed. Baby Sophie had a stormy passage, but she survived. She came out of hospital three weeks later with irreversible brain damage.

Over the years, I have gone over the whole scenario many times. Could I have done anything differently and achieved a different outcome? I don't think so. It was clear that Mary didn't think so either. For the next ten years Mary brought Sophie to see me every three months, without fail. I might tweak her anti-convulsant medication as she grew, but other than that, there was not much medical that happened. It was more about the human connection that was forged on that day, and the support I could give to Mary as she dealt, admirably, with the cards that life had dealt her.

8.2 1990–1999: INDEPENDENCE

Life as a village GP was hugely satisfying yet challenging. Starting a family of my own meant that our family life was often interrupted by the needs of my patients. It was hard for my wife, who raised our two children effectively on her own on the weekends or nights when I was on call. When I was out, it was she who had to answer the phone to patients. She would bleep my pager and I would call her back to get the details.

The dependent relationship between doctors and patients had been this way for decades in the practice. In truth, it worked well for both parties, most of the time. As we went into the 1990s, the impact of Margaret Thatcher's neoliberal economic policy took its toll. The demands on both me and my family increased. The rise of consumerism, with its inherent selfishness and greed meant that, for me at least, the model became increasingly unsustainable.

'I've just come for an MRI scan. I've got headaches'. That was his opening line. I knew him as a rather pompous local business owner. I tried to get a better idea as to what was happening, and what might be the cause of his headaches.

But he was having none of it. 'I've told you; I've got headaches. Had them for weeks. I just need an MRI scan to rule out a brain tumour'.

I explained, getting exasperated myself, that I could not, as a GP, request an MRI brain scan. That would need to come from a neurologist. First, I would need to take a history and examine him to establish the most probable cause. He was quick to circumvent that suggestion.

'Fine. Just refer me to a neurologist, then'.

The consultation ended badly. Both of us were angry and dissatisfied. A few weeks later I received the neurologist's report with the result of the brain scan. He had chronic sinusitis. I felt vindicated, and slightly gloating. But that is not how I *wanted* to be. That is not the doctor I had trained to become.

I hardened. The increasing demands of patients had meant I had to harden my own boundaries. Seven and a half minutes per consultation (eight consults an hour) is not long to deal with presenting problems. We never quite got round to putting up notices that said ONE PROBLEM PER CONSULTATION, but several nearby practices did.

I am sure that she must have seen me look at my watch as she mentioned her third problem. It did not take long to resolve. Hardly warranted mentioning at all. I was well over time as she walked to the door.

But then she paused and said, 'I wonder if I should mention the lump in my breast?'

My heart sank. I wish I could say it was with empathy for the sudden realisation that this was the real reason that she had come to see me. But I am ashamed to say it was my own thought: 'I really can't deal with this right now'. What I did not realise then was that this was one of the first signs of becoming burnout.

The resentment got worse. One Saturday evening I got a call from a woman who told me that her father had swallowed his false teeth (presumably just a partial rather than a complete set). It was 7:30 p.m. We were settling the kids.

I started to take the history. 'When did this happen?'

'I think it was last Saturday. We were just chatting after dinner, and he realised he hasn't seen them yet. You know, they haven't . . . come through.'

This could have been the start of a funny story. But I did not see the funny side. My personal and family boundaries were not being respected. I felt angry. I am sure I was civil, but I was not the caring doctor I wanted to be.

I shared this with my senior partner in the practice. He was the most caring doctor I have ever met, but we were caring to the extent of depriving ourselves for the sake of our patients. We had a healthy debate about an opinion article in the *BMJ*, called 'The caring doctor is an oxymoron' (Short, 1998). I had come to believe that it was possible for doctors to care *too* much; that in seeking to obtain independence and assert their rights, patients continued to expect that the responsibility for their health rested with the doctor.

This, I now realise, is the root cause of doctor burnout. Independence could have worked, but only if patients were overtly responsible for their own health.

In this phase of the developing independence between doctors and patients, the definition of health remained as it was before. Health was the absence of disease. But health had become a commodity that patients were seeking from doctors. The absence of responsibility from this commoditisation was appealing, I'm sure, but it was never going to work. It led to patients growing dissatisfied with their doctors. Complaints increased. The doctor–patient relationship had become tense and fractious.

Since that time, the conflict in the relationship has continued to grow. The COVID-19 pandemic has accelerated the enmity. Recently, a colleague of mine wrote on LinkedIn that he had been phoning his patients, asking if they wanted advice about the safety of vaccination.

One gave him a two-word response over the phone: '*F@#k off!*'

The locus of control has shifted. Asserting self-authority of this nature is an unhealthy form of the internal locus of control. This radical independence means that the relationship is fractured. Yes, doctors will continue to do their job, but the humanity which has been integral to the role of the doctor for centuries will struggle to survive.

Over the years, I had increasingly done more work in occupational health during my time in general practice. I obtained several higher qualifications in occupational medicine. So it was with interest that I received an invitation to dinner with two directors of an occupational health company. By the end of the evening, they invited me to leave general practice and train to become an occupational physician, a specialist doctor in the health of workers.

8.3 2000 TO THE PRESENT: INTERDEPENDENCE

In living memory, no one had left the practice unless they were retiring. There was surprise and disappointment when I told my patients that I was leaving. My partners were supportive and accepting. They kindly threw a leaving party for me to which patients were invited. Many of my old favourites came. I am humbled even now as I recall the speeches and the cards and gifts. It felt like the sort of event that should occur at the *end* of a long and illustrious career, not part-way through.

A week after the leaving party, I called into the local petrol station. I went in to pay. It was quiet – I was the only customer. I knew the man behind the counter.

'Ah, Dr Beaumont, just the man! I've got a lump in my groin. I think it's a hernia. Can I show it you quickly?'

He was not one of my old favourites. He had not been at the leaving party.

'Oh, I'm sorry, I've left the practice. That wouldn't be appropriate.'

As I left, I felt gleeful and empowered. I had just said 'No' to a patient.

I may have grown weary and cynical, but I had played my part in how we had reached this stage of how it worked between us, doctor, and patient. My humanity was fading in a way that makes me cringe as I recall it.

Thank goodness for occupational medicine and the company that headhunted me out of general practice. Soon afterwards a feature article about me was published in the *BMJ*, with my reflections about entering a career in occupational medicine, under the heading, 'My dream job'. And it was.

My relationship with patients shifted with immediate effect. I did not see them in a clinical setting: instead, I was seeing them at their place of work. They had a job to do for their employer. I was engaged by their employer to help keep them safe from the hazards of work. There was shared responsibility. I had to give correct professional advice; the employer had a legal duty to keep them safe and, in turn, the workers had a legal responsibility to follow whatever rules were in place to keep them safe.

They stopped being patients. They were now *clients*. I was no longer Dr Beaumont, I was 'Doc'. But it was used with a degree of irreverence (and endearment), as in, 'Who is your next victim, Doc, who do you want to see next?'

My other role was rehabilitation and return to work for injured workers or workers with a medical or psychological condition. What could we do to support them to come back to work? It was not my job to treat them, or to provide any interventions (by that time I was no longer prescribing medication of any kind). My job was to work out why they had not returned to work when they would have been expected to. Why had a man with a fractured ankle not returned to work after six months?

My new specialist training in occupational medicine provided the framework to understand what was happening.

The biopsychosocial model was first described in 1977, by Dr George Engel (Engel, 1977). It never ceases to astound me how few people have encountered it, including doctors. It builds on the medical model (biomedical model), which says that if there are symptoms, they are caused by underlying pathology which can be identified and treated (or not). But the biomedical model is too simplistic to account for the man with a fractured ankle who has not returned to work after six months. The fracture has long since healed, but he still has chronic pain. Why? The biopsychosocial model says that beyond the physical factors (the 'bio' of the name), there are psychological and social factors at play. The longer that someone is off work, the more likely that psychosocial factors become more significant than the physical ones.

Psychosocial factors can be addressed, of course. But usually by the time an occupational physician is involved, the combination of factors is so complex that only a team approach will overcome the various barriers to recovery. The team must be multidisciplinary, typically comprising a physiotherapist, an occupational therapist, and a psychologist, together with the doctor.

When I try to explain chronic pain to someone, a common response is, 'Wait a minute, you're saying it's all in my head?' It took me a while to feel brave enough to say, 'Yes, I am' – although I am usually gentler than that.

I explain it like this: If you tread on a drawing pin, where do you feel the pain? Not in your foot. The nerve impulses travel from your foot to your brain, where they are interpreted as pain. All pain, of whatever nature or cause, is always 'in your head'.

In chronic pain, even after the injury or pathology has healed, a process has been set up in the brain whereby even normal sensations coming from that region are experienced as pain, and often as severe pain. Known as 'central sensitisation', this is a primitive defence mechanism whereby the brain seeks to protect you, long after the threat has disappeared.

There's good news and bad. The bad news is that often no treatment will cure central sensitisation. Painkillers are poor at controlling this kind of chronic pain. Interventions like surgery can do nothing because there is no injury or pathology left to treat. The good news is that, caught early enough, the brain can be reassured that the threat has passed, that the brain no longer needs to protect the body. Everything is OK and the person can recover. It is the job of the multidisciplinary team to help the person understand their pain and develop the techniques to heal themselves. This is the point at which responsibility needs to shift, so that the locus of control is with the patient, not with the medical team.

If I am good at my job as a physician, I bring all my learning, experience, and clinical acumen into play. There is great satisfaction in working out all the factors at play. But this is an intrinsic motivation. The satisfaction that I feel is my own. I do not need anyone else to reward me or to tell me I have done a good job.

And the client? Their job is to recognise that the responsibility for what happens to them is theirs alone. Success or failure in recovery and return to work depends on the client playing their part and practising the techniques that the team has shown them. To understand that the control of what happens to them rests in their own hands.

This is a big shift from everything we were taught about the relationship between 'doctors' and 'patients' or, as I prefer to term it, between healthcare and people. For some people this is a concept beyond their grasp. They cannot move from an external locus of control.

'Well, that didn't work! What are you going to do next?'

This situation, which is encountered frequently, probably by all doctors, is what I now see as the root cause of burnout in doctors. As one colleague put it recently, 'The problem is that patients expect me to fix them'.

Doctors do not fix patients. They help create the circumstances for patients to heal themselves. There are a few circumstances where a specific surgical treatment may effect cure or restore function, for instance certain cancer surgeries or joint replacement surgery, but in most cases, whatever a doctor does or prescribes is simply aiding the body's own amazing capacity for healing or recovery. Antibiotics support the body's own immune system to fight infection. Suturing a wound brings together the skin edges so that the wound can heal. Antidepressants enable the mind to regain control of mood.

The role of antidepressants is now less well-understood than we thought it was 20 years ago. There is considerable evidence that a major mechanism of action of efficacy for antidepressants is the placebo effect. The placebo effect is one of the most amazing examples of how the body can heal itself – that simply by believing that a drug is going to work, it does. Even if (as in drug trials) that 'drug' is a sugar pill. For me, this was a watershed in my understanding of the relationship between doctors and patients (or me and my clients).

Before, the doctor's role was to tell people what to do, taking responsibility for their health outcomes. Afterwards, the doctor evolved into an expert in understanding health and supporting people to understand their own health, to enable them to take control of their life and health.

Once the doctor has helped create the circumstances for healing (working with a team and, preferably, operating within a supportive environment and system), the person is able to take responsibility for their healing, their recovery, and their return to function. In my professional work, one explicit goal is a return to work. In the situations I usually encounter, 'return to work' is a proxy measure for returning to life.

I have had my own brush with mortality and adversity. At the age of 42, overweight and unfit, but with no other risk factors, I suffered a heart attack. This put life into perspective for me. Suddenly and unexpectedly, the doctor had become a patient.

It is important for the patient to play their part in the medical process and be the recipient of medical expertise – on the understanding that medical science is only creating the circumstances for the heart to heal. I made a good recovery – and I take a cocktail of preventative drugs to this day.

This was my first realisation that health is more than merely physical.

My Chief Medical Officer came to see me.

'David, I'm really pleased you've made such a good recovery, but I'm worried about your state of mind.'

That was my first encounter with depression. I had felt that I was at the pinnacle of my career, and suddenly it had all been taken away. It took a psychologist to help me realise that this was only my own fear and limiting belief. I picked up the pieces and slowly, gingerly, resumed my search for the truth about health and the role of doctors.

That search took me to New Zealand. My brush with mortality enabled me and the family to decide that a move of 12,000 miles across hemispheres was just one big adventure.

The next few years were exciting and challenging, professionally, and personally. In New Zealand, occupational medicine was different: less time in the workplace, more in the clinic. The focus was on injury management and on identifying and removing barriers for the client's return to work.

I saw and assessed thousands of people who had *not* returned to work, who had become stuck in their recovery process. I came to see them as having been let down by the system. At that time, the system was very much geared towards physical diagnosis and on whether ACC, the Accident Compensation Corporation, New Zealand's injury compensation legislation, could be applied. Psychosocial factors were often salient, but it was hard to take them into account in terms of the compensation system. This led to anger, frustration, and contention with the clients.

For me, too, psychosocial factors became salient. I was middle-aged, still overweight, and unfit. Our marriage was struggling and in time we parted. Depression returned. Before antidepressants and counselling came my dark night of the soul, my descent to rock bottom, and my realisation that there had to be more to life than this.

My strict religious upbringing had led me to rebel as a teenager. I progressed to nihilistic, atheist beliefs. Suddenly it seemed laughable to me that I had spent my whole life devoutly believing in *nothing*. My journey to an exploration of belief systems and spirituality had begun.

If there is a purpose to life, what was *my* purpose? It became clear to me that my purpose is to change the concept of health.

Suddenly health was more than merely physical, more than psychological. Health encompassed relationships, emotional health, and spiritual health. In fact, my own experience taught me that health pervades every area of life. I came to see that my clients were expressing and requesting my acknowledgement of this broad concept of health which hitherto had not been acknowledged.

Two things happened. The first was that I became involved with the Royal Australasian College of Physicians and, within the College, with my own professional body, the Australasian Faculty of Occupational and Environmental Medicine. The second was that I authored a book to share what I had learned (Beaumont, 2021).

If health pervades all areas of life, then it follows that the focus of my speciality, work, is important.

There are two parts to the job of the occupational physician: protecting workers from the hazards of work and helping them recover from illness or injury so they can get back to work.

But something was missing.

After I joined the Policy and Advocacy Committee of my Faculty, we began a piece of work to explore the relationship between work and health. We found that, while some work can harm you, much work is beneficial. We wrote a position statement, *Realising the Health Benefits of Good Work*. The subsequent consensus statement was met with wide agreement and signed by many organisations throughout New Zealand and Australia, representing workers, unions, employers, doctors, other healthcare professionals, and the government.

We found common ground. Work can have a positive effect on people. Working, doing good work, improves health. 'Health' encompasses psychological, emotional, and spiritual well-being – all the things that give meaning and purpose to life.

As I began to research the concept of health, I discovered that the old definition was being challenged and debated. The work led by Professor Machteld Huber in the Netherlands seemed particularly resonant. For Huber, health was 'the ability to adapt and self-manage in the face of social, physical, and emotional challenges' (Huber et al., 2011).

Later, I simplified Professor Huber's concept. Health is the ability to adapt and self-manage; in short, health is the ability to take control of one's life.

Huber saw health being an integral part of every aspect of our lives. It is a positive state, rather than the absence of disease. She proposed the concept of 'positive health' (Huber et al., 2011).

Like many doctors of my generation, I was not taught about the origins of health, and had never encountered the concept in my professional development. Yet as far back as 1979, the year I had entered medical school, the US-Israeli medical sociologist Aaron Antonovsky was writing about salutogenesis: the origins of health (Antonovsky, 1979).

Having understood that health is a positive construct and pervades every area of our lives, I began to realise that the ancient wisdom of Indigenous people around the world have always known this. There are holistic, or whole person models of health in so many different cultures. In New Zealand, the medical profession uses the Māori model of health,[1] described and named by Professor Sir Mason Durie in 1984. He called it, 'Te Whare Tapa Whā', literally 'the four sides of the house of being'. The solid house that makes up the healthy self: physical health, psychological health, whānau (family) or relationship health, and spiritual health.

So much had coalesced for me. For years I knew that I had a book to write. I had made three false starts. Now I took six weeks off work and the book poured out of me. The only limit to how fast the pages were written was the speed I could type.

If the practice of medicine were to change, such that doctors were empowering people to have positive health, what would that practice be called? Positive Medicine (Beaumont, 2021). I titled the book, *Positive Medicine: Disrupting the Future of Medical Practice*. Yes, disruption, but *positive* disruption.

The framework I used for the discussion was Te Whare Tapa Whā. I was honoured to have this approach endorsed by Sir Mason Durie himself, who wrote the foreword to the book.

The proudest moment of my professional career was to have the acceptance of Oxford University Press to publish the book, following peer review. Later, when I discussed the book with the Regional Manager of a major New Zealand construction company, Fulton Hogan, he said, 'I see you use positive medicine for individual clients, but could you develop a group programme for my workers? They need this right now'.

---

[1] Mānatu Hora: Ministry of Health. Māori health models – Te Whare Tapa Whā www.health.govt.nz/our-work/populations/maori-health/maori-health-models/maori-health-models-te-whare-tapa-wha [accessed 12 November 2022].

We were just coming out of the first lockdown of the COVID-19 pandemic in New Zealand. There was fear and uncertainty. A feeling of loss of control, in an unfamiliar new world. Those workers, like all of us, needed to gain control of their lives, so they could attain positive health.

I worked with a pilot group of workers, hand-picked to not hold back. They did not. Half an hour into the very first session, one of them interrupted me and said, 'Well, this isn't going to work'. My heart sank, but I sensed from the mood in the room, and even how I felt, that he was right.

'Why not?' I asked. He told me.

'What would work?' He told me.

'OK. Let us stop right there and meet up again in three weeks. I will re-write the programme.'

The principal change I made was to take them on a journey. If the four pillars of health of the Māori model of health was the framework, then the process was the hero's journey. I was helping them believe that they could become the heroes of their own life journey, not passive participants in a world that was shaped by the pandemic. I was helping them change their locus of control from being acted upon to acting. Health is the ability to control our life. I was helping them to define what positive health meant for them.

I called the programme 'Project Me: Be the Project Manager of Your Own Life'. Unlike the paternalistic role I had held as a GP in my early years, I no longer needed to tell people what to do. They knew already. The answers are always within. A key part of the process was giving them time and space to stop and think about themselves. To reflect on what they each needed to have positive health and well-being.

Well-being: The term originated in positive psychology, the science of human potential and flourishing.

There are many definitions of well-being, which tend to be narrative descriptions because of the nature of the concept. One that I favour is as follows:

> The combination of feeling good and functioning well; The experience of positive emotions such as happiness and contentment as well as the development of one's potential, having some control over one's life, having a sense of purpose, and experiencing positive relationships.
>
> **(Huppert, 2009)**

My own working definition is that well-being is living life to our full potential and flourishing.

My client, the Regional Manager, was so pleased at the response and engagement of his team that he asked me to design a programme for the community in which his workers live. He said, 'I would like to give back to the community'. The programme had already opened for partners and families to attend. Now we were impacting the health of the whole community.

My role as a doctor has changed from the one for which I was trained. So, too, have my relationships with the people whom I serve. Once an expert, I have become more of a coach or mentor. I use my specialist expertise, knowledge, and life experience to help my clients take responsibility for their own health and their own lives.

Although I had seen positive medicine as being a model of medical *practice*, I came to see it as more than that: a process that enables people to find positive health and well-being.

Along the way, I have learned a good deal from inspirational colleagues. One such is Dr Sam Hazledine. He is committed to improving the lives of doctors. He has helped many who have become burnout, as I was when I left general practice. He sees the altruism of doctors as a potential weakness. Often, doctors do not prioritise their own health and well-being. The evidence is that doctors with impaired health provide inferior care to their patients. So, he decided to do something about it.

Sam Hazledine convinced the World Medical Association that doctor burnout was such a prominent issue that it must be included in the Physician's Oath of the Declaration of Geneva, the modern equivalent of the Hippocratic Oath. As a result of his submissions, the oath now includes the following statement: 'I will attend to my own health, wellbeing, and abilities in order to provide care of the highest standard' (Hazledine, 2017).

The pandemic has taken a huge toll on all healthcare workers. This prompted our team to develop a programme specifically designed for doctors. It was tested for the first time at the 2022 Congress of the Royal Australasian College of Physicians (RACP). At the start of the 90-minute session, participants were asked what words they would use to describe their health. This was collated electronically into a word cloud. Some people used the words 'happy' or 'grateful'. But the most common words were 'suboptimal'. Others said they were stressed, tired, undervalued, frustrated, and anxious.

By the end of the session, the most common word was 'hopeful'. Others said they felt calmer, empowered, motivated, optimistic, peaceful, and purposeful.

Of course, this is not conclusive on its own, but it certainly indicates a shift in mindset and a change in the locus of control. We do not know whether the change will be sustained. But here is at least the start of a process to support the health and well-being of doctors, and to help them realise that their own health is so much more than the absence of disease.

Fifty years after deciding I was going to be a doctor; I have arrived at a vastly different understanding of what health is and how doctors should be engaged in the delivery of health. The concept of health I see now is much broader and more positive. It continues to evolve with feedback.

I believe that by changing the concept of health, and the way that people interact with doctors and the healthcare system, the focus can shift from treatment of disease, beyond prevention of disease, to enhancement of health and well-being – living life to our full potential and flourishing. This should apply just as much for doctors and healthcare professionals as the people they serve.

The doctor of the future will:

- Prioritise their own health and well-being to be able to give the best care to the people they serve.
- Be trained both in pathogenesis, the origins of disease, and salutogenesis, the origins of health.

- Be trained in the application of positive medicine into their practice.
- Be expert in both disease management and positive health.
- Empower people to take responsibility for their own health, with an internal locus of control.
- Be trained in a spectrum of practice from disease management to health enhancement, requiring differing skills from technical expertise to coaching and mentoring. Choosing their careers according to their own aptitudes and aspirations in relation to these varying modes of practice.
- Contribute to the health and well-being of communities and populations, adding to overall levels of happiness, with the goal of a flourishing society.

REFERENCES

Antonovsky, A. (1979). *Health, stress, and coping*. Jossey-Bass.

Beaumont, D. (2021). *Positive medicine: Disrupting the future of medical practice*. Oxford University Press. https://doi.org/10.1093/oso/9780192845184.001.0001

Engel, G. L. (1977). The need for a new medical model: A challenge for biomedicine. *Science (New York, N.Y.)*, *196*(4286), 129–136. https://doi.org/10.1126/science.847460

Hazledine, S. (2017). *Medicine's four-minute mile: The evolved physician's oath – 9 steps to achieve wellbeing*. Medworld.

Huber, M., Knottnerus, J. A., Green, L., Horst, H. van der, Jadad, A. R., Kromhout, D., Leonard, B., Lorig, K., Loureiro, M. I., Meer, J. W. M. van der, Schnabel, P., Smith, R., van Weel, C., & Smid, H. (2011). How should we define health? *BMJ*, *343*(7817), d4163–d4237. https://doi.org/10.1136/bmj.d4163

Huppert, F. A. (2009). Psychological well-being: Evidence regarding its causes and consequences. *Applied Psychology: Health & Well-Being*, *1*(2), 137–164. https://doi.org/10.1111/j.1758-0854.2009.01008.x

Short, A. D. (1998). The caring doctor is an oxymoron. General practice will develop best if "caring" is replaced by professionalism. *BMJ (Clinical Research Ed.)*, *316*(7134), 866–867. https://doi.org/10.1136/bmj.316.7134.866a

9 Building Health Resilience and Capacity through Functional Medicine Nutrition

My Journey as a Disenfranchised Biomedical Scientist

Shelly Cavezza

9.1 THE ROLE OF BIOMEDICAL SCIENCE IN INDIGENOUS HEALTH: LESSONS LEARNED

I have always been an opportunist and I am open to experiencing new things and going with the natural flow of life. In my early 20s, while travelling in Australia, I met a friend who offered me a laboratory position at a small community college and, thus, my science career began.

The first few years were fun, and I enjoyed the friendly, busy role of a laboratory technician. However, as I learned and improved my skills on the job, I realised that I wanted to do something more meaningful. So, while engaged in full-time employment, I completed an undergraduate science degree and, later in my late 20s, I completed my PhD in Biomedical Science.

I undertook My PhD to address the problem of scabies in remote Indigenous communities in northern Australia. In Australia, scabies affects about six in ten Aboriginal and Torres Strait Islander children, more than six times the rate seen in the rest of the developed world (Clucas et al., 2008; Cuningham et al., 2019). In these remote communities, community members commonly accept multiple infections during childhood and they accept chronic illness later in life as part of life. This is in part due to the multi-generational impact of the disruptive colonial history of Australia but also due to low socio-economic status, a lack of access to fresh affordable food, and limited housing availability leading to overcrowding.

Scabies was, and still is, endemic in most of these remote communities and most of the dogs in the community have scabietic mange (Walton & Currie, 2007). A microscopic mite burrows into the epidermis of the skin and lays eggs and causes

DOI: 10.1201/9781032711195-11

scabies. It generates relentless itching and a significant rash, and often the skin gets secondarily infected with bacteria. In these remote communities, babies and young children are treated for multiple scabies and bacterial skin infections before the age of 2. At the time of my PhD, there was a big push for a limited allocation of government health dollars spent on treating the mangy dogs to protect the children from infection.

My PhD thesis was the first molecular work in the world on the scabies mite with the overarching aim to develop new diagnostics, treatments, and/or vaccines. The focus was to genetically identify and clone some of the allergenic proteins secreted by the mite causing the itchy skin reaction. This work was initially unsuccessful but after several years of perseverance and some serendipity, we published the first studies on the genetic epidemiology of scabies mites (Walton et al., 1997). My studies identified that people were getting scabies primarily from other people, not their dogs (Walton et al., 1999). This had a direct benefit to the communities because it meant that control programmes for human scabies in endemic areas did not require resources directed against zoonotic infection from dogs (and the scarce government funding spent on treating mangy dogs could be reappropriated).

Although my work was rewarding, the longer I worked on the molecular aspects of scabies, the more marginalised I felt in the biomedical science community. I always wanted to do translational research, where the outcomes of the research were of direct benefit to the individual. It became increasingly apparent that, in the eyes of the biomedical research community, success meant the high total dollar value of successful grants and the number of papers that were published by academics. It almost felt like the actual proposed project outcomes were of secondary interest. It was, and still is, an extremely competitive environment and, at the time, it felt like no one seemed to care if you finished the project or translated the learnings. In the early 1990s, Indigenous health was not high on the political agenda. Thankfully, this has begun to change. Of course, the white elephant in the room was that no pharmaceutical company was ever going to support the development of a new vaccine/drug for scabies due to the low return on investment for any new therapeutics in this area. Globally, scabies affects low socio-economic populations, including displaced people living in migrant camps, who can ill afford to pay for a new drug just released on the market.

9.1.1 Understanding that Early Childhood Events Can Be the Triggers of Chronic Ill Health Later in Life

Scabies is regularly the precursor to impetigo (school sores) and severe skin and blood infections in childhood (Steer et al., 2009). These bacterial infections have been linked to rheumatic fever, acute post-streptococcal glomerulonephritis (APSGN), and kidney disease (Thomas et al., 2015). Outbreaks of APSGN usually coincide with scabies outbreaks, which can contribute to the development of chronic kidney disease and subsequent renal failure in adulthood. So, there was and still is every reason to eliminate this parasitic skin disease from children living in affected communities.

When this system breaks down, instead of health, humans can experience core clinical imbalances that may cross many different organ systems long before overt disease is recognised or diagnosable. Our health history is important. Often, we forget or don't realise that the infections or traumas experienced in our early life can present as chronic diseases later in life.

When the body processes environmental inputs such as infections, food, stress, exercise, and toxins, it affects our metabolism. Our genetic makeup, physiology, spiritual and emotional beliefs, and attitudes also impacts on our metabolism.

9.1.2 Conventional Western Medicine versus Traditional Medicine

My interest in natural medicines began during my time researching scabies in northern Australian Indigenous communities. Tea tree oil is a natural bush medicine used for many years in remote Australia for the treatment of scabies and skin sores. We were able to demonstrate the effectiveness of tea tree oil and its active agents to kill scabies mites in the laboratory, as well as its antibacterial and anti-inflammatory effects (Walton et al., 2004). This work opened my eyes to the other traditional medicines that are often ignored in conventional medicine. This is due to the fact community healers and those adept at telling health stories often describe the beneficial effects of traditional herbal medicine in anecdotal reports. Communities pass down these learnings, but the scientific journals may not publish these, like the double-blind placebo-controlled clinical trials accepted as gold standard by the scientific community. This early work also led to later collaborations with Chinese immunologists to investigate the Chinese traditional medicine Lei Gong Teng (*Tripterygium wilfordii* Hook F.) and its immunomodulatory and anti-inflammatory effects for rheumatoid arthritis. The good news is that the World Health Organization is now supporting studies in scientific validation of traditional medicines to treat many chronic and infectious diseases (World Health Organisation [WHO], 2022)

9.1.3 Food as Medicine

Another significant insight I gained during my studies on scabies occurred when working with people with a very distressful and confronting disease known as crusted scabies. In crusted scabies, the immune system is unable to control the mite burden and the skin becomes extremely thickened and infested with millions and millions of mites. These people become core transmitters in the community and often become stigmatised, isolated, and rejected by their families. My postdoctoral work focused on understanding the genetic and immunopathogenesis of this disease and how we could potentially support the immune system to move it back to a more protective response. One factor that became apparent in my research was that some of the precursors to poor health, impaired immunity, and disease susceptibility were poor diet, nutritional deficiencies, and social and emotional isolation and rejection (Walton & Weir, 2012).

During my time at the research institute, another group was working on improving nutrition in remote communities. Their goal was to support the integration of bush medicine back into the diet and provide a nutritionally dense alternative to the highly processed fast food available at the bush store. It was only after I left that I appreciated the value of their work and the importance of this approach to individual immunity and overall community health.

Studies have shown that traditional bush foods include the consumption of lean meats and a large variety of mostly raw plant foods rich in fibre and complex carbohydrates, such as fruits, vegetables, seeds, nuts, and roots with a low energy density but a high density of some nutrients and antioxidants (O'Dea, 1991). More recent research shows that phytochemicals, various biologically active compounds found in plants, work as a family of nutrients that serve as immune modulators through their ability to favourably influence cellular signalling systems (Bland, 2021a). These phytonutrients interact with the intestinal microbiome to help modulate the systemic immune system and may help support immune system renewal. In other words, they develop regenerative and agile pathways to help the body construct health.

9.2 THE ROLE OF TEACHER AND RESEARCHER IN ACADEMIA: A PERSONAL EVOLUTION

After resigning from my position as Senior Biomedical Research Scientist, I took up a position as Associate Professor of Immunology at Australian University. This position opened my eyes to the highly competitive university environment and the insidious effects of corporate culture on the individual.

Highly demanding and overwhelming workloads, staff not feeling valued in the workspace, and the relentless stress resulted in my diminishing personal health. The doctors diagnosed me with autoimmune diseases (coeliac and psoriasis) and long-term symptoms of arthralgia, poor sleep, low mood, allergies, and food/chemical intolerances.

I felt constrained by the incessant procedural and protocol-driven landscape. My creativity diminished drastically, and I felt hopelessness about the future. This, I later recognised, was a clear driver of ill health. During this period in my life, I became dissatisfied with the standard medical interventions that were also offered to me with the focus on 'a pill for an ill', e.g., immunosuppressants, antidepressants, and anti-inflammatories.[1] It started to feel like I needed pills and medications to go to work and perform in my job!

This emphasis on medication and/or surgery to resolve health problems instigates a sense of disempowerment for the individual and reliance on the health practitioner for a quick fix for the health issue. Practitioners in turn, must undermine the quality of their recommendations due to time and cost constraints. The therapeutic relationship is poorly mediated by a diagnosis with a prescription pad. Patients are wanting

[1] These, of course, just shut down or reduce the symptoms but do not address the triggers or cause of the symptoms, nor do they facilitate the trajectory in the body to build new pathways to health.

critical thinking and mindful involvement with a whole-of-life understanding by their practitioners.

We now know a fair amount about the adverse physiological effects that stress has on health (e.g., it raises cortisol levels), but the connections in the body go many ways. Stress can increase the need for certain nutrients, but certain nutrients can also mitigate the physical symptoms and subjective response to acute stress. My knowledge of the immune system and its complex, intricate, and redundant network of molecules, cells, tissues, and organs also made me start to rethink the 'biologics' and anti-cytokine therapies appearing on the market. I reflected on the 'Butterfly Effect' and what were the longer-term consequences of this type of therapy. The Butterfly Effect was a term coined in the 1960s by Edward Lorenz, a meteorology professor at the Massachusetts Institute of Technology, who was studying weather patterns. He proposed that small, seemingly trivial events may result in something with much larger consequences – in other words, they have non-linear impacts on complex systems.

> Web-like interconnections between mind and body and back again move us out of the single agent–single outcome way of thinking. Instead, the body is an interconnected whole organism within which everything affects everything, and nothing is truly isolated.

9.2.1 A Unique Way to Practice and Focus on Healthcare

What if there was a different way? I became increasingly drawn towards natural therapies and traditional medicines.

I started attending conferences and workshops that discussed evidence-based natural therapy and nutritional approaches to health, health prevention, and building immune resilience. Health practitioners have become well-versed in the signs and symptoms of diseases, but are we able to observe the signs of optimal balance and wellness? How do we 'diagnose' positive vitality?

I discovered and found my tribe in Functional Medicine – a systematic whole-body approach to health incorporating nutritional and lifestyle strategies aiming to support, reset, and optimise the body and mind (Bland, 2021b, 2022). Understanding key subclinical imbalances and their potential effects on the individual is one way to see health. However, extending this further is the understanding that key essential and non-essential nutrients need to be at optimal levels for the presence of vitality, not just the absence of disease. Vitality also includes resilience and the ability of the body to withstand challenges that may arise from genetic predispositions, infectious diseases (colds, flu, COVID), increased physical or emotional stress, harmful environmental and toxin exposures, or dietary changes. Therefore, the body also needs reserves to draw on when challenged.

To this end, I enrolled in postgraduate studies in Functional Medicine, Human Nutrition, and Nutrigenomics (the relationship among nutrients, diet, and gene expression) and started applying the principles to my health. I changed my diet (after listening to my body and how it reacted to certain foods). I addressed my nutritional deficiencies and insufficiencies (identified through blood biomarker testing). I improved my gut microbial dysbiosis (a result of many years of being unaware that I was gluten intolerant), and I included counselling, yoga, and slow relaxing walks in my mental health plan. The outcome was a noticeable improvement in my energy levels, health biomarkers, and increased vitality. However, it wasn't until after I left the university and my stress levels decreased that my health started to improve, and my creativity returned.

My mother experienced a lifetime of stress and early childhood loss and trauma which significantly impacted her health and well-being and resulted in her diagnosis of bipolar and early dementia in her late 50s. She stood out to me as an example of what my future might look like if I continued the same trajectory with a high-stress load. She trained and worked as a nurse for many years during my childhood but refused nearly all medication to treat her illness. This at the time was frustrating but I now see her lack of trust in the conventional medical approach. Her faith sustained her and although times got very tough for her, she always fought for her right to choose. She became a guide for me to follow a more natural and nutritional way of achieving strength and well-being.

9.3 THE ROLE OF HEALER AND ADVOCATE IN FUNCTIONAL MEDICINE NUTRITION: COMING HOME

In 2019, I resigned from my university position and started a clinical practice consulting in Functional Medicine Nutrition integrating the latest health, nutrition, and wellness research into personalised health management plans for clients. My goal became to improve health outcomes for individuals who were struggling or disappointed with their current conventional medicine healthcare plan. My goals were to use my skills and expertise towards optimising the health and wellness of my clients through the assimilation of cutting-edge scientific evidence, traditional medicines, and individual needs.

9.3.1 Identifying and Addressing the Causes of Ill Health (Not the Symptoms) Is Key to Improving Well-Being

One client that comes to mind is Nick,[2] a 19-year-old student who came to see me with chronic long-term sinus congestion, ear, and chest infections; asthma; allergies; and constipation since childhood. He had just finished a 20-day course of antibiotics, the 8th course of antibiotics in that year, and he had residual congestion and inflammation. As you can imagine, Nick was exhausted, with the only option of more antibiotics offered to him by his general practitioner. The constant sinus congestion, chest infections, and lack of sleep were affecting his quality of life and ability to concentrate on his studies. The numerous courses of antibiotics over his lifetime had impacted his gut microbiome and ability to digest and absorb nutrients, as well as

[2] Not his real name.

impairing his immunity. He also had developed chronic lower back, foot, and knee pain, a clear sign that the inflammation was now systemic. Nick was desperate for an alternative approach.

We set to work to identify and address the key triggers and mediators of his symptoms including diet, food sensitivities (dairy, gluten, and corn), allergies (his mum's cat and house dust mite), gut dysbiosis (malnutrition/maldigestion/malabsorption), and stress. We focused on improving his diet by switching from a processed and fast-food diet to organic whole foods removing known inflammatory, mucous-forming foods. We also provided immune-modulating and gut-repair supplements to support his digestive and immune systems. Other interventions included keeping the cat out of his bedroom, nasal flushing, reducing mobile phone use at night, 15 minutes of sun exposure on waking, and 20–30 minutes of daily walking for the anti-inflammatory effects and to reduce stress load.

After four weeks on the health plan, Nick's gut was 80% better, but he was still having sinus and sleep issues. After eight weeks on the health plan, his gut was 90% better, and his sinuses were great, with no further infections or colds in the prior two months (the first time in years!). After 12 weeks, his symptom score had reduced to 20, from an initial 120. His sinuses were still good despite a recent exposure to the cat and pollen. He had no further colds and flu infections (three months in total) and his energy levels were great.

When people experience long-term symptoms, they are often related to chronic inflammation, sometimes of unknown origin. A lengthy period of declining function in one or more of the body's systems often precedes chronic disease. Returning patients to health requires reversing (or improving) the specific dysfunctions that contributed to the disease state. Those dysfunctions are, for each of us, the result of lifelong interactions among our environment, our lifestyle, and our genetic predispositions.

9.3.2 The Interconnected Web of the Human Body, Mind, and Environment

Using the model of a web can help explain health in which one major imbalance can influence many distinct functions and systems. If there are no major imbalances, but several pathways are off-centre, the overall pattern of health will be affected. A web is complex and interlinked – multiple factors can influence a single condition, and multiple conditions can affect a single dysfunctional process or imbalanced system. Rather than looking for a single 'root cause' of a disease and finding the 'cure' in a single pill, functional medicine works to identify what is out of balance, what has skewed the stream of biochemical information, energy, and emotions too far to one side or the other of a healthy range. For multiple conditions, especially those associated with ageing – like Alzheimer's disease, heart disease, or autoimmune conditions – it is often not possible to find a single root cause. No single problem exists, and no single pill will cure these conditions in every person. Instead, there are likely numerous triggers and imbalances, each contributing something to the outcome.

Unravelling the web will find a myriad of factors with an extraordinarily complex set of interrelationships for each of these conditions (Jones, 2010).

For example, a hypersensitive/dysregulated immune system mixed with exposure to dietary and environmental antigens sets the stage for the clinical diagnosis of 'allergy'. Allergy symptoms can include a wide range of immune-mediated adverse reactions such as rhino-conjunctivitis, chronic sinusitis, dermatitis, epilepsy, migraine, hypertension, joint inflammation, and mental depression. The immunopathogenesis includes multiple mechanisms including IgE antibodies and histamine. Our understanding of 'allergy' and other chronic diseases must also include awareness of the interconnections among food, intestinal flora, and systemic health as well as contributions from neurogenic inflammation. Changes in the gastrointestinal microbiome can provoke a cascade of physiologic responses that may lead or contribute to widespread imbalances and result in a variety of symptoms that may or may not conform to a recognised pattern or named disease.

Individualised assessment, nutritional optimisation, healthy lifestyle, environment interventions, and treatment of dysbiotic bacteria (whether in the gut, genitourinary tract, or nasopharynx) provide the hope of a cure rather than an endless and additive round of antibiotics and anti-inflammatory drugs.

9.3.3 Why Do We Age Differently?

Looking at people in general and particularly within my own family, there are noticeably clear differences in ageing, with my mother's health declining rapidly after the age of 65, whereas my father is still living a healthy independent life at age 92.

In Functional Medicine, the assumption is that if we can reduce oxidative stress and inflammation among patients and identify and address their genetic susceptibilities, we will mitigate adverse environmental impacts and stress. In turn, micronutrients will replete while consumers develop healthy lifestyles, thus creating resilience and adaptability in their bodies. This enables people to live out their years with longer periods of health and vitality and shorter periods of disease and debility. It is now becoming possible with new epigenetic testing to obtain information on biological age (epigenetic age) as opposed to chronological age, including steps on how to reduce overall ageing.

Currently, the best biomarker tests of an individual's biological age are based on patterns of DNA methylation. DNA methylation is the addition of a methyl group to cytosine residues at selective areas on a chromosome. Of 20+ million methylation sites on the human genome, there are a few thousand which are tightly correlated with age. These epigenetic tests can assess the effects of preventive measures on the reserve capacity of our body to respond to life's insults – illness, trauma, loss, chronic stress, and poor lifestyles.

A recently published randomised controlled clinical trial demonstrated that an 8-week treatment programme that included diet, sleep, exercise, and relaxation guidance, and supplemental probiotics and phytonutrients was able to reduce biological ageing. Forty-three healthy adult males aged 50–72 were able to reduce their age on average by 1.96 years by the end of the 8-week programme (Fitzgerald et al., 2021) – that translates to three months of chronological age reduction for every week of programme participation (a 1-week investment to achieve a 12-week age reduction!)

The Functional Medicine Nutrition model now offers some hope to those who have lost confidence in the conventional allopathic model of health, the system in which doctors and healthcare professionals 'treat diseases and symptoms' versus the approach that considers a whole-of-person and whole-of-life ecological and evolutionary health model. It is also useful for those who want to achieve a healthy lifespan and who are not willing to accept disability later in life as a 'natural part of aging'. It provides a new model of healthcare practice that addresses the drivers of chronic diseases such as cardiovascular disease, diabetes, obesity, cancer, autoimmune and allergic conditions; and digestive, mood, hormonal, and cognitive disorders by translating new and existing research into a systems biology approach. The model is based on a strong client–practitioner relationship, and the development of a collaborative health plan incorporating individual capacity, genetics, diet, nutrition, environmental exposures, stress, exercise, and emotional and mental health needs. Each person is biologically unique, and one size never fits all. Assessment and recommendations first address the patient's core clinical imbalances, fundamental physiological processes, environmental inputs, and genetic predispositions, rather than heading straight for the diagnosis.

9.4 SUMMARY

Our bodies have an amazing ability to self-heal. We can restore balance in a dysfunctional system by strengthening the fundamental physiological processes that underlie it and adjusting the environmental inputs that nurture or impair it. This chapter explained the professional and health journey of a biomedical scientist who continues to learn and marvel at the wonders of regenerative self-healing made possible through a holistic, Functional Medicine approach.

9.5 KEY POINTS

1. We are each genetically, biochemically, and environmentally unique, with our own unique health history.
2. Functional Medicine offers a personalised healthcare plan that treats the individual. It supports the normal healing mechanisms of the body, addressing the triggers of ill health naturally, rather than just masking the symptoms.
3. Our bodies are a complex web or network of both internal and external systems and interconnections. Understanding the connections allows us to see deep into the functioning of the body.
4. Listen to the whispers. Our bodies are intelligent and have the capacity for self-regulation to achieve dynamic balance and resilience when we meet nutritional, emotional, and environmental needs.
5. Health is not just the absence of disease, but a state of balance and positive vitality which is able to enhance the health span, not just the life span, of each individual.

9.6 RECOMMENDATIONS

- Find and work with a healthcare practitioner who provides a patient-centred approach (not disease-centred) to treatment.

REFERENCES

Bland, J. S. (2021a). Application of phytochemicals in immune disorders: Their roles beyond antioxidants. *Integrative Medicine (Encinitas)*, *20*(5), 16–21. www.ncbi.nlm.nih.gov/pubmed/34803535

Bland, J. S. (2021b). Treatment adherence, compliance, and the success of integrative functional medicine. *Integrative Medicine (Encinitas)*, *20*(3), 66–67. www.ncbi.nlm.nih.gov/pubmed/34373681

Bland, J. S. (2022). Functional medicine past, present, and future. *Integrative Medicine (Encinitas)*, *21*(2), 22–26. www.ncbi.nlm.nih.gov/pubmed/35698609

Clucas, D. B., Carville, K. S., Connors, C., Currie, B. J., Carapetis, J. R., & Andrews, R. M. (2008). Disease burden and health-care clinic attendances for young children in remote Aboriginal communities of northern Australia. *Bulletin of the World Health Organization*, *86*(4), 275–281. https://doi.org/10.2471/blt.07.043034

Cuningham, W., McVernon, J., Lydeamore, M. J., Andrews, R. M., Carapetis, J., Kearns, T., Clucas, D., Dhurrkay, R. G., Tong, S. Y. C., & Campbell, P. T. (2019). High burden of infectious disease and antibiotic use in early life in Australian Aboriginal communities. *Australian and New Zealand Journal of Public Health*, *43*(2), 149–155. https://doi.org/10.1111/1753-6405.12876

Fitzgerald, K. N., Hodges, R., Hanes, D., Stack, E., Cheishvili, D., Szyf, M., Henkel, J., Twedt, M. W., Giannopoulou, D., Herdell, J., Logan, S., & Bradley, R. (2021). Potential reversal of epigenetic age using a diet and lifestyle intervention: A pilot randomized clinical trial. *Aging (Albany NY)*, *13*(7), 9419–9432. https://doi.org/10.18632/aging.202913

Jones, D. S. (2010). *Textbook of functional medicine*. Institute for Functional Medicine.

O'Dea, K. (1991). Traditional diet and food preferences of Australian Aboriginal hunter-gatherers. *Philosophical Transactions of the Royal Society London B Biological Sciences*, *334*(1270), 233–240. discussion 240–231. https://doi.org/10.1098/rstb.1991.0112

Steer, A. C., Jenney, A. W., Kado, J., Batzloff, M. R., La Vincente, S., Waqatakirewa, L., Mulholland, E. K., & Carapetis, J. R. (2009). High burden of impetigo and scabies in a tropical country. *PLOS Neglected Tropical Diseases*, *3*(6), e467. https://doi.org/10.1371/journal.pntd.0000467

Thomas, J., Peterson, G. M., Walton, S. F., Carson, C. F., Naunton, M., & Baby, K. E. (2015). Scabies: An ancient global disease with a need for new therapies. *BMC Infectious Diseases*, *15*, 250. https://doi.org/10.1186/s12879-015-0983-z

Walton, S. F., Choy, J. L., Bonson, A., Valle, A., McBroom, J., Taplin, D., Arlian, L., Mathews, J. D., Currie, B., & Kemp, D. J. (1999). Genetically distinct dog-derived and human-derived Sarcoptes scabiei in scabies-endemic communities in northern Australia. The *American Journal of Tropical Medicine and Hygiene*, *61*(4), 542–547. https://doi.org/10.4269/ajtmh.1999.61.542

Walton, S. F., & Currie, B. J. (2007). Problems in diagnosing scabies, a global disease in human and animal populations. *Clinical Microbiology Reviews*, *20*(2), 268–279. https://doi.org/10.1128/CMR.00042-06

Walton, S. F., Currie, B. J., & Kemp, D. J. (1997). A DNA fingerprinting system for the ectoparasite Sarcoptes scabiei. *Molecular and Biochemical Parasitology*, *85*(2), 187–196. https://doi.org/10.1016/s0166-6851(96)02825-3

Walton, S. F., McKinnon, M., Pizzutto, S., Dougall, A., Williams, E., & Currie, B. J. (2004). Acaricidal activity of Melaleuca alternifolia (tea tree) oil: In vitro sensitivity of sarcoptes scabiei var hominis to terpinen-4-ol. *Archives of Dermatology*, *140*(5), 563–566. https://doi.org/10.1001/archderm.140.5.563

Walton, S. F., & Weir, C. (2012). The interplay between diet and emerging allergy: What can we learn from Indigenous Australians? *International Reviews of Immunology*, *31*(3), 184–201. https://doi.org/10.3109/08830185.2012.667180

World Health Organisation (WHO). (2022, March 25). *WHO establishes the global centre for traditional medicine in India*. WHO. www.who.int/news/item/25-03-2022-who-establishes-the-global-centre-for-traditional-medicine-in-india

10 From Rehabilitation to Prevention and Work Design

The Journey of a Healthcare Practitioner

Bharati Jajoo

10.1 PRESENT DAY

> *'Ma'am, my neck and back pain has increased: I am not able to sit and work.'*
>
> *'Our employees are complaining about their new desks and chairs. We recently invested in building a state-of-the-art facility, yet they continue to complain. We are working with change management teams to resolve the concerns, but is there anything you can help us with?'*
>
> *'We purchased new chairs, but employees still complain about their aches and pains!'*
>
> *'We want to do ergonomic study for our product assembly line!'*
>
> *'We are building a new floor. Do you have any suggestions for the design of our workstations?'*
>
> *'We have a new prominent and important client, and we want to highlight our line of products. We are competing with global players. How can we leverage your knowledge to educate clients about the ergonomic benefits of our seating systems to consider our product line?'*

These are some of my professional interactions with organisational clients, individuals, and equipment suppliers who require my services as a registered occupational therapist and ergonomic assessment specialist, working in Bengaluru (Bangalore), India.

10.2 HOW DID THIS START?

I started working in the occupational therapy department of a large public hospital in Mumbai, India, where I had the privilege to work in hospital-based inpatient and outpatient units to rehabilitate individuals with injuries, illness, disablement, and

DOI: 10.1201/9781032711195-12

disease. We serviced 'patients' rather than 'clients', like I do now (this orientation matters) (Ramdass et al., 2008; Ratnapalan, 2009; Reilly, 1984). So, our patients were from all walks of life from slum areas to those who worked as highly trained professionals. As a master student in the same institution, our work postings rotated every month to different units, ranging from psychiatry, plastic surgery, hand surgery, medicine, orthopaedic, neurology, cardiology, and more. Every patient had different stories to tell, sometimes lengthy, and we had to extract information to determine what was relevant to their rehabilitation. We recognised that the intimacy by which they shared their stories revealed parts of their being and we needed to consider this to help them become their most capable and independent selves.

My admission to occupational therapy in central India's biggest teaching hospital, established in Nagpur, India, was merely an accidental phenomenon, or an intentional mishap, which changed the trajectory of my life and shaped my world view. As is typical in India, after 12 years of your basic education, family and educators encourage you to pursue further education. Until I applied to occupational therapy, I was not aware that such a profession existed. My friends were applying, and they happened to have an extra application form: I, too, applied (my angel sang this into truth).

Many years later, I am so thankful for that momentary decision to submit that application. Our undergraduate studies in a medical college gave us grand exposure to all basic medical subjects. The hands-on approaches with a senior mentor gave us exposure to patient intervention. Coupled with therapy, we learned to build required aides, modifications, splints, braces, or whatever patient needed to facilitate their independence in their daily living tasks, e.g., self-care, job tasks, or transportation. In some cases, this meant helping families and caregivers support the person while exerting the least effort using principles of joint protection, work simplification, and energy conservation.

As a student, I did not fully appreciate the impact on prevention that an occupational therapist could make with their professional skills: our focus was primarily on rehabilitation. This model concentrated on maximising the functional potential of people after their experience of injury, illness, or disablement, to be a productive member of society who found meaning in life. I now understand what it means to orient practice in a person-centred way (Cloninger, 2010), considering the 'patient' as a 'client', and asking them about their therapy goals. When the practice is driven by the client goals and they are fully engaged in the vision of potential change and the pathway to attain that experience, a resilient pathway is forged (Rodríguez-Bailón et al., 2022). We must determine nuanced, customised care approaches that consider the person's personality, needs, or culture. We can share the goals of rehabilitation and, if so, foster mutual respect. Our empathy and reflective practice will lead to system improvements. The goals and outcomes can be more realistic when collaborative approaches to health promotion are undertaken (Cloninger, 2010), and the occupational therapist can better balance the triad of wants, needs, and the realities of the person receiving care and their circumstances.

During our oncology department postings, we worked with terminally ill patients, helping to mobilise them to aid their basic self-care. Sometimes, we fabricated a splint or a brace to position their body parts in optimal way to reduce their pain or improve their function. During this time, it was challenging to stifle my concerns and knowledge about the medical condition and knowing that the clinical outcome was terminal. In the beginning as a junior student, I might have wondered, 'What is the point?' With experience, and as well-articulated by Dr Atul Gawande in his descriptions about 'being mortal', when he explains that life is meaningful because it is a story, a story has a sense of 'whole' and 'essence' with a thread of purpose and, in stories, endings matter (Stolarek, 2016). Certainly, suffering at the end of life can be unavoidable, but so too we suffer throughout life. Helping people find their grace throughout this journey is an altruistic goal. As students in our emerging practice, we might not have fully realised the important of this, but this recognition of the need to dignify all stages of life was a developing concept. The medical settings, in which I stationed my early work, considered people as 'patients' but my understanding of client-centred care with purposeful and meaningful engagement in activity as a hallmark of health was evolving.

10.3 MOVING TO A FOREIGN COUNTRY

In the early 1990s, many therapists from India and from other countries dreamt of working in the Western world: the United States or the United Kingdom, for example. When we finished our master's programme in occupational therapy, most of my colleagues and I appeared for the professional registration exam to be licensed to work in the United States. As soon as we received our results, we received calls from recruiters.[1] I ignored them because I appreciated that my conservative family would not likely permit me to travel to the United States alone for work. In my community, as soon as studies are concluded (and sometimes earlier), the family's prominent concerns are how to ensure their child is wed and 'settled'. Dreaming about work choices were not the focus.

However, one late evening, my dad was visiting me from my hometown and heard a recruiter's message. He asked about it. I explained the opportunities to work in the United States and, to my surprise, he was elated that his daughter, the first generation of women in the family line to complete a degree, had an opportunity to work in a foreign country and that this was in the United States, no less. His stipulation was that I had to travel with a known and valued friend. Luckily for me, the hospital offered my classmates and close friends jobs in the same hospital. And so, the adventure began.

My friend and I landed in Philadelphia. Our conscientious and generous department director collected us from the airport. Although I understood that December was very cold in Philadelphia, and we brought the warmest available winter jackets from Mumbai, we froze! That Mumbai 'heavy jacket' did not feel like more than a simple cotton shirt in the wintry weather. The director was insightful and somehow predicted that we might not have been well prepared: she had two extra jackets ready for us upon our arrival.

As we drove towards our apartment, I looked at the quiet highway in the dark, in the middle of the night, driving from the airport to our new home. What I saw out

[1] Quite often in the middle of the night in India.

from the window was speeding cars and barren trees. As the headlights fell upon these trees, just that sight without seeing many people, so much quietness, cold, and uninhabited in those moments, made me feel incredibly homesick. My stomach filled with butterflies. In that moment, I wished that the car could turn back, and I would return home on the return flight to Mumbai, on the same plane in which I had travelled. Freedom and adventure were a prison of anxiety to me momentarily.

The next day, we visited the occupational therapy department at our new hospital-based workplace. A senior therapist invited us to shadow her for a few days and then we progressed to seeing our own caseload of people awaiting their therapy. It was an unusual practice for us to introduce ourselves first and ensure that we received the patient's permission to observe their therapy engagement while collaborating with a senior therapist. And there was our first hurdle: our accent. People could hardly understand us, even despite speaking as slowly as we could, clearly, enunciating our words (or so we thought). Soon after our employment start, we enrolled in speech therapy classes to effectively neutralise our accents so that verbal communication was no longer a barrier when working with hospital inpatients and outpatients, and our colleagues.

I genuinely enjoyed my outpatient posting. This is when I first learned about work hardening and what it entails to facilitate return to work for injured workers. A functional activity analysis was useful to consider readiness for work hardening (Matheson et al., 1985). As such, we had to understand the job demands and simulate the frequent or high-demand work tasks in a graduated way. There were instances when I shadowed the senior therapist to attend a worksite to better understand these job demands, since there were no documented profiles in those instances. However, our focus was still on the individual and their return to work, not on design or preventive and proactive system-wide measures.

Ten months passed quickly, and my work visa would expire soon. However, circumstances had changed on my personal front, and I was engaged to be married. My fiancé was on the West Coast, so I looked for jobs in this region. I was successful in my application at an orthopaedic doctor's private practice. This was an intimate clinical set-up, and doctors mostly referred their clients to see me because of their upper-extremity functional needs. In our therapy protocol, we used modalities and pain management techniques. We provided home exercise programmes and instructional aide to support daily living activities. Sometimes, this involved the prescription of and the advice in the use of functional aids to prop work or living tasks. Here, also, we simulated work tasks and inquired about critical job tasks, though we remained in the clinical environment, not the industrial one. We provided feedback to the orthopaedic surgeons who decided on the client's return to work capacity based on our advice.

10.4 A PLACE WHERE MY ERGONOMICS KNOW-HOW MATURED

A year after my work in the West Coast of the United States, I had another opportunity to work in a hospital, though this time it was a larger establishment. I was hired as the first dedicated full-time outpatient occupational therapist, and I worked in a multidisciplinary team environment.

Hospital jobs were not easy to come by, especially in Northern California, so I embraced this opportunity. On my first day, after a brief orientation, the senior occupational therapist handed over several outpatient client cases and said, 'These are your babies now. Also, here are two work hardening clients'. And this was where and how I began. The difference in this setting was that we had a dedicated space with more specialised and supportive equipment than I had seen before. I feigned my knowledge of these tools and supplies, and I researched feverishly to understand their use. However, before the era of smart mobile device use, on-the-fly research was more difficult. I gathered my courage to discuss these shortcomings with the senior therapist and asked for assistance, demonstrations, and explanations around their protocol of use. With patience, I learned to expand my scope of service. I found that my colleagues, each in their specialty areas, were enthusiastic about their area of work and we shared our knowledge to enhance clinical care delivery.

A client, Julie,[2] came to see me, and she was one of the nurses from within the hospital. It was her third presentation in as many months with ill effects from a recurring injury. My colleague asked if I could help improve her health status by placing her in work hardening. I asked about her work tasks, critical physical job demands, the decisions that she made, and her opinions about her job and the organisation of her work. In the clinic, I reviewed her functional movement patterns and body mechanics, and noted areas for improvement with corrective techniques. Soon, her treating physician released her with full clearance to her substantive job duties.

A few days later, I covered for a colleague who was away, and I was on the ward floor visiting an inpatient. I saw Julie in action at her work. She was not under my care, but I paused to observe the realities of her work (Rayner, 2023). It was a far cry from the interview disclosures or our clinic observations and simulations in controlled quarters. In the actual functions, Julie often worked in cramped places without adequate space for good body positioning. In this instance, I understood that Julie was unable to comply with any clinic-based recommendations for her posture, duration, or repetition exposures during her tasks: the work setting, and job demands did not permit her actualisation of rehabilitation strategies.

I realised then that our clinical assessments missed too many factors to adequately inform effective workplace rehabilitation. In the healthcare environment, when distractions are common and work situations constantly change (Rathert et al., 2009; Sharma et al., 2016), it is imperative to design for this variability. As such, my recommendations must address work system design (Dul et al., 2012), not just personal health strategies adopted periodically within an imperfect system.

Unsurprisingly, Julie returned to the outpatient clinic, injured again, and her complaints caused her to miss several more days at work in her nursing role. At the same time, I noted that our occupational therapy outpatient caseload had increased, often backfilled by referrals for people returning with recurring medical complaints. The gender and job role of the person varied, male, female, different job title, but similar concerns about recurring injuries. They either could not return to the work, or they had to modify and restrict their work tasks because of lingering musculoskeletal or

[2] Fictious names have been used in this submission to ensure client confidentiality.

comingled psychosocial complaints, like fatigue. I administered quality treatment confined to the scope of the clinic-based intervention. The clients improved, but then, like a revolving door syndrome, several would return after months back at work, with symptoms exacerbated or a new raft of related symptomology. Some clients worsened with their work exposures, requiring surgical intervention (such as to relieve nerve entrapment). We provided the post-operative rehabilitation, like mobilisation techniques, scar management, nerve gliding, stretching, strengthening, and instructions to modify and graduate exposures to home or work activities.

One of the clinic's medical record management staff members presented with upper quadrant symptoms of discomfort. When advice was provided, she agreed that the recommendations would be ideal, but that the constraints of her restricted work area made it impossible to adopt. 'I have no luxury to execute what you advise', she said, 'Could you visit my workplace?' Well, that was not in our service regime; we were clinic based. However, it sounded like a sensible progression to inform meaningful recommendations. Since we worked in the same hospital facility, I visited her at lunch time.

I witnessed a multitude of contributing factors that affected her health and productivity and I understand them as 'ergonomic risk factors'. So, without understanding the formal linkages to ergonomics processes at the time, my foundations of occupational science led me to understand the rationale underpinning site observations to assess performance capability.

Subsequently, I probed at length to learn more about the realities of work among my hospital-based colleagues. We had so many employees attending therapy for their injuries and occupational rehabilitation needs. There were, simultaneously, staffing needs and the pressure to return staff to their substantive job roles. The employee health nurse visited our department to inquire about several employees and their return-to-work status. There were clinical patterns among the cohorts of workers and their presentation of injuries, yet the pressure was mounting to manage these cases and ensure an early return to work (for financial reasons because of insurance costs but also because of the staffing shortages).

I made the opportunity to include ergonomics training in my continuing development training. One of the most favourable aspects about practising occupational therapy in the United States was the opportunity to invest in continuing education. It was more than 'permitted', it was a requirement for licensure and registration. I leveraged my past educational experiences from courses like physical agent modalities, shoulder rehabilitation, stroke rehabilitation, women's health and wellness, and manual therapies, to make sense of some of the education about ergonomics. However, I needed more than the science and clinical aspects of ergonomics, I needed the business and design aspects: e.g., how to document the assessment and justify the findings and recommendations. I needed to leverage the findings to inform organisational change or job redesign, communicate this in an effective manner to decision-makers, and collaborate with workers to discover the realities of their work and devise design strategies to mitigate risk factors and construct better health and working conditions.

My education in ergonomics proved to be the most valuable addition to the repertoire of my problem-solving skills as an occupational therapist. It provided me with the overarching framework to consider a systematic and scientific way to evaluate work systems and injury causation. It helped me communicate what was previously instinctive by formalising my discovery approaches. Notably, third party reimbursement parties recognised the service code, 'ergonomics', in 2001,[3] well after my trajectory of expansive service delivery in this realm. Since it was not a recognised service category during my early practice, there were times when I had to be creative when conducting my evaluations. The constraints and coordination of on-site observations were significant, and my role confined me to conducting visits for hospital employees, not other outpatients (I could not justify the visits to their place of employment as a reimbursable service. The hospital employees and their work areas were accessible to me).

During our case management review meetings, our employee health department noted the increasing worker's compensation costs and the work-related cumulative trauma cases were significant. The hospital employees were presenting from all departments: housekeeping, cafeteria, microbiology, pathology, pharmacology, nursing, medical specialist staff, and intensive care unit staff. I used an allocated treatment time to visit their work area, workspace, and the contributing risk factors that impacted on their health and performance. It was clear to me that behavioural-based personal health intervention strategies were not going to resolve the concerns: we needed to do more in terms of a holistic, systems-based design strategy.

While my awareness grew about ergonomic risk factors, the solutions and simple accommodations were sometimes difficult to resolve. Ergonomic strategy sessions were not part of our standard service protocol. I relied on my early education about fabricating splints and making adaptive aids for rehabilitation patients in Mumbai. My emerging practice aligned with my foundations in occupational therapy because the focus was on ensuring that people could engage in meaningful functional tasks.

During a client case review meeting, I presented the challenges that I had discovered by using an ergonomics lens during the work design discovery (Karanikas et al., 2021) while observing our hospital employees at work. Our employee health management teams recognised that there were many pieces of the proverbial puzzle when designing a safe and productive work environment for the hospital workers. A decision was made to establish an 'Ergonomics Task Force', and I was invited to coordinate this task force, reporting to the employee health department. Our task force team was comprised of an employee health nurse and representatives of human resources, purchasing, and hospital operations. We invited two injured employees to provide their perspective about the risk exposures and work experiences. Our mission was to establish an initiative-taking, preventive ergonomic programme for hospital workers while we developed new processes for workspace development (like new fit outs or refurbishments), equipment, or product procurement.

We reviewed injury record-keeping data, identified trends, and noted high-risk work departments. I focused on completing job and task analysis to understand the physical, cognitive, psychosocial, environmental, and organisational context of work in the hospital. I identified and catalogued the risk factors and began to consider tasks that could be readily modified or redesigned without much capital investment

[3] www.federalregister.gov/documents/2000/11/14/00-28854/ergonomics-program)

and made the plans or prepared the business cases when major capital expenditure was needed. We categorised our solutions into three principal areas: engineering solutions, safe work practices or personal protective devices, and administrative solutions. We also developed teaching and training modules for every department based on our finding. From this, we promoted the process of early reporting while education employees about risk factors, hazard identification, and early warning signs of discomfort or symptoms. We also worked with the employee occupational health nurse to accommodate workers with modified duties and suitable work plans to enable them to remain in their work role and use work as a medium of their rehabilitation (Matheson et al., 1985). These aspects forged our path towards preventive and proactive work design.

10.5 REALISING THE RESULTS OF THE ERGONOMICS TASK FORCE

Over the next three years, or ergonomics programming and work processes evolved and matured. Ergonomics education became a central part of the employee induction, and it was a module in our Yearly Education Services ('YES') programme. The hospital management consulted our task force on design strategy when they scheduled new refurbishments, or when they procured capital equipment. In short, we were becoming part of the fabric of business operations, and there was a recognition of the important of ergonomics in work design in most operational aspects of the hospital.

The hospital reaped the business results, because of their investment in ergonomics. The worker's compensation claims reduced after several years, and more employees sought early intervention help for work-related discomforts, rather than waiting for serious injuries to arise. While we could have measured more outcomes to profile these successes, at this time I had no knowledge of the ways in which to value proposition of ergonomics. However, anecdotally, I loved watching the level of employee engagement rise, because workers saw that we valued their work and their experiences at work. We did not formally measure this at the time, but it was evident to me in my interactions with my colleagues.

10.6 MY THOUGHTS TODAY

I returned to Bengaluru (Bangalore), India, and I work as an ergonomics practitioner. I focus on ergonomics in design, whether it is job crafting, work design, office set-up, environmental design, product design, or establishing corporate health and well-being and ergonomics programmes. Ergonomics is the scientific study of people at work and the systems of work (Wilson, 2012), and it involves a comprehensive and integrated analysis of examining tasks, workspaces, controls, displays, tools, equipment, products, supplies. lighting, training, work systems, and governance, to understand the needs of a diverse array of people in and affected by the system. I still struggle with advancing the value proposition of ergonomics because it is not widely known or understood. My academic colleagues publish scientific papers, but we need more business case studies published in influential business media. We must make ergonomics accessible and readily understood so that the business and consumers can embrace these practices. Once you apply an ergonomics lens, it is impossible to 'unsee' the world from the view of design improvements that involve human actors. That is, if you have a business objective, and humans must advance that objective, then design should focus on the humans who must perform well. Healthcare and other organisations must consider ergonomics as a 'must have', not a 'should' or 'could' have. Our healthcare workers are critical parts of our social systems and we must design well for their well-being and performance. I will continue to do my part to advance these ideas and I hope that by sharing my story, more decision-makers will be influenced to embrace the ideology of good work design with human-centred practices.

REFERENCES

Cloninger, C. R. (2010). Person-centred integrative care. *Journal of Evaluation in Clinical Practice*, *17*(2), 371–372. https://doi.org/10.1111/j.1365-2753.2010.01583.x

Dul, J., Bruder, R., Buckle, P., Carayon, P., Falzon, P., Marras, W. S., Wilson, J. R., & van der Doelen, B. (2012). A strategy for human factors/ergonomics: Developing the discipline and profession. *Ergonomics*, *55*(4), 377–395. https://doi.org/10.1080/00140139.2012.661087

Karanikas, N., Pazell, S., Wright, A., & Crawford, E. (2021, July 25–29). Advances in ergonomics in design, proceedings of the AHFE 2021 virtual conference on ergonomics in design, USA. *Lecture Notes in Networks and Systems*, 904–911. https://doi.org/10.1007/978-3-030-79760-7_108

Matheson, L. N., Ogden, L. D., Violette, K., & Schultz, K. (1985). Work hardening: Occupational therapy in industrial rehabilitation. *The American Journal of Occupational Therapy: Official Publication of the American Occupational Therapy Association*, *39*(5), 314–321. https://doi.org/10.5014/ajot.39.5.314

Ramdass, M. J., Naraynsingh, V., Maharaj, D., Badloo, K., Teelucksingh, S., & Perry, A. (2008). Question of "patients" versus "clients": "Patients" versus "clients". *Journal of Quality in Clinical Practice*, *21*(1–2), 14–15. https://doi.org/10.1111/j.1440-1762.2001.00396.pp.x

Rathert, C., Ishqaidef, G., & May, D. R. (2009). Improving work environments in health care. *Health Care Management Review*, *34*(4), 334–343. https://doi.org/10.1097/hmr.0b013e3181abce2b

Ratnapalan, S. (2009). Shades of grey: Patient versus client. *Canadian Medical Association Journal*, *180*(4), 472–472. https://doi.org/10.1503/cmaj.081694

Rayner, A. (2023). Improving systems: Understanding "work as imagined" versus "work as done". *In Practice*, *45*(1), 55–57. https://doi.org/10.1002/inpr.282

Reilly, M. (1984). The importance of the client versus patient issue for occupational therapy. *The American Journal of Occupational Therapy*, *38*(6), 404–406. https://doi.org/10.5014/ajot.38.6.404

Rodríguez-Bailón, M., López-González, L., & Merchán-Baeza, J. A. (2022). Client-centred practice in occupational therapy after stroke: A systematic review. *Scandinavian Journal of Occupational Therapy*, *29*(2), 89–103. https://doi.org/10.1080/11038128.2020.1856181

Sharma, K., Hastings, S. E., Suter, E., & Bloom, J. (2016). Variability of staffing and staff mix across acute care units in Alberta, Canada. *Human Resources for Health*, *14*(1), 74. https://doi.org/10.1186/s12960-016-0172-1

Stolarek, I. (2016). Dr Atul Gawande on being mortal: Does it matter? *Australasian Journal on Ageing*, *35*(1), 62–62. https://doi.org/10.1111/ajag.12309

Wilson, J. R. (2012). Fundamentals of systems ergonomics. *Work*, *41*, 3861–3868. https://doi.org/10.3233/wor-2012-0093-3861

11 Health and Wellness for Those 65 Years and Older

Tim Henwood

For older adults, the value of exercise participation is irrefutable and backed by reams of research evidence dating back to the 1970s. A simple search of the National Library of Medicine PubMed website using the term 'benefits + exercise + older adults' reveals close to 16,000 peer-reviewed publications supporting the value of participation (National Library of Medicine, Search date January 20, 2023), with over 14,000 of these studies published since 2000. Breaking this down further, there are over 750 systematic reviews in this space and more than 1,000 publications specific to the value of exercise in the residential aged care (nursing home) setting. Looking at just one of these, Sherrington et al. (2019) undertook a systematic review of 108 randomised controlled trials with 23,407 community-dwelling participants (age 76 years; 77% women) and reported that exercise reduced the rate and number of falls among older adults with the benefits most prominent when programmes included balance, functional exercises, and resistance training. In contrast, among older Australians, falls are the leading contributor of hospital admission and the leading cause of injurious death. One in three adults over 65 years of age fall each year, with the falls rate per person increasing each year after an individual's first fall. Among older women who fracture their hip from a fall (hip fractures account for 26% of all fractures and nine out of ten injurious fall hospitalisations), 40% will never walk again, 33% will be forced into a residential aged care setting, and 24% will die within a year of the fall. In Australia, 41% of all hospital spending ($3.6 billion) was dedicated to injurious falls in 2015–2016.[2] But what is most amazing is that while the dollar invested in falls prevention continues to increase, the percentage of older adults (number of falls/total population over 65 years) and deleterious impacts of falls have remained the same over the past 30 years and even longer.

Falls prevention is one example of exercise being an underutilised and under-prescribed treatment pathway for older adults. This extends to every disease, frailty, sarcopenia (age-related loss of muscle mass and strength), and loss of independence. It is said that good research takes a while to translate to practice. However, this timeline is unbelievably slow and, given the implication of population ageing to the individual, their loved ones, and to health and aged care spending, it is unforgivable. More must be done by way of access to quality exercise prescription and motivational pathways or environments conducive to exercise for the older adult.

DOI: 10.1201/9781032711195-13

The above-mentioned 16,000 research publications about the relationship between exercise and healthy ageing cover a plethora of interventions, timelines, cohorts, and measures, but all tell the same story: exercise works. We are not just talking about getting fitter because exercise positively influences everything! Of course, like younger adults who engage in physical activity, when older adults positively participate in regular and ongoing exercise, their aerobic endurance improves, their muscles get bigger, and their bones get stronger (United Kingdom National Health Service, 2021). Before we move on to the host of other benefits, let's pause and look at bone. The impact of bone loss with increasing age is the primary factor (fracture accounts for 50% of all hospital admissions) in the depressing story told above about falls. However, where osteoporosis has been a long-acknowledged issue, the loss of muscle has only more recently gained attention with the frailty and sarcopenia revolution. Why is this an unbelievable oversight? You can have low bone mass or bone quality and still live independently, but when you have low muscle mass or muscle quality, independence becomes near impossible. More frustrating is that we have also long understood that positive bone remodelling is associated with the regularity and intensity of muscle contraction. Sit on your backside all day and limit activity and your bones get smaller and more brittle but challenge your muscles and your bones get bigger. Even the medication and supplements we are encouraged to take don't work without the stimulus of exercise to promote bone growth.

Exercise reduces the risk and the symptoms of every disease. This extends to terminal and palliative cases, where it can reduce associated pain and fatigue. In the mental health space, it has the same impact, and can prevent mental health disease development and reduce the risk of disease progression. So much so, there is solid evidence that it is as effective as medication to alleviate depression and anxiety. You could even say that it is more effective than medication when we consider the risk of polypharmacy – the medication side effects that range from nausea to constipation or dizziness. In addition, it is strongly advised that once you are on antidepressants, you should never go off them. *But that's ok*, because there are other pills you can take to counter the side effects of your antidepressants! This impact of medication side effects is far reaching and goes well beyond antidepressants, aligning directly to the increasing numbers of older adults on medications to counter their increasing number of chronic diseases. Eighty percent of all older Australians have at least one chronic condition, with 39% having three or more. This latter group has an 80% increased risk of mortality when compared to those with no chronic diseases. In addition, they visit their GP at least monthly, their specialist five times a year, and their pharmacist at least every eight days. Medical visits and medications become the mainstay of their lives, with some medications to counter their primary disease and others to counter the side effects of their primary medications. This is because half of all patients with multiple chronic health conditions have conflicting treatment regimens that complicate disease management (Gilbert et al., 2013). Polypharmacological patients with $\geqslant$5 medications we call 'Rattlers'. An example would be someone with four different disease diagnoses and four primary medications to counter these, then six more secondary medications to counter the side effects of medication one to four. As you would appreciate, life doesn't get easier with more pills. Sides effects, misadventure (when you take it at the wrong time, for example), and prescription errors are all common, with Australian data suggesting that one in three unplanned hospital admissions among older adults are medication related.[3] However, when exercise is used to alleviate depression (or any disease for that matter) the side effects are better health and well-being! Common side effects include bigger muscles and bones, reduced falls and fracture risk, and a reduced need for other medication because exercise can alleviate the symptoms of disease. And we still haven't mentioned the impact of exercise on mental health, cognition, or social and community engagement. Further, that it allows older people to push back on disability, that it increases quality of life, that it improves continence, and the big one, it is the only stimulant that gives a glimmer of hope to a shared person-centred goal, the ability to remain independent, living in our own homes for as long as possible, i.e., until we die.

The World Health Organization's and Department of Health Promotion life course model explains that the maintenance of the highest functional capacity with increasing age is a measure of well-being and ageing (WHO, 2000). In simple terms, it breaks life into three stages: Early, Adult, and Older Life. The model gives two alternative trajectories to follow. One crashes through the disability threshold in older life where the other does not. Your lifestyle choices determine which of these trajectories you are on. Poor diet, sedentary behaviours, smoking, drinking too much, poor sleep hygiene, or stress, and 'Boom!' you can be subject to bad health and an earlier check-in to dependent care. Choose the healthy pathway and you can push back on disability long into later life and, in contrast, to a slow but guaranteed fade away. Per the health-oriented pathway, people will experience a compressed period of mobility or, even better, die quietly during sleep in their own bed. This group is the anomaly though. However, a body of work Dr Anthony Tuckett and I did a while ago, informed by the Australian & New Zealand eCohort Study of Nurses and Midwives, showed that if you picked two healthy lifestyle behaviours, you were better off than someone who only adhered to one, and three better than two, four better than three, etc (Tuckett & Henwood, 2014). For the clients accessing our services, the team of allied health professionals, therapists, and carers help clients push away from dependent care and back towards the better health alternative. We do this by offering clients holistically planned, meaningful, physical, social, and mental health engagement, and we constantly reinforce better health literacy to ensure clients appreciate why they are at the gym, why we are increasing their weights and modifying their programmes, and why we are referring them to our exercise physiologist, physiotherapist, occupational therapist, dietician, podiatrist, social worker, and/or registered nurse. At the forefront to all this is the alignment of the client's assessed needs to their person-centred goals. And whether these goals are short- or long-term, complex like a hiking tour of Nepal, or simple like still being part of a family gathering, the underlying theme is always the same, they all want to remain independent for as long as possible.

As an Exercise Gerontologist working with older adults now for over 20 years, both in the academic and the aged care setting, I have engaged in many client goal-setting activities, and the underlying goal statements rarely vary. 'I just want to stay in my own home. Not one of these nursing facilities or villages! They're for old people!' Or some version of this. (I started a section here, 'If I had a dollar for every time I had heard' but rapidly realised I would be lucky to have $50,000. I am good at what I do, but not 1 million people good . . . yet). So many people want the same outcome, and so few people achieve this. In Australia, over a quarter of a million older adults are hospitalised each year following a fall, over 1 million are using aged care supported services for showering and/or domestic assistance, and 5% of all older Australians over 65 years and 18% of the 80 years plus group reside in residential aged care. Residential aged care has improved significantly in Australia over the past few decades with improvement in care and food standards creating hotel-like experiences in some cases. Providers have also become more proactive and innovative about their models of care. Complementing this, the Aged Care Quality Standard changed in 2019, moving the provider off the podium and replacing them with the client. This placed the client fairly and squarely where they should be, as the most important thing in their aged care journey. Even though the evidence is clear (overwhelming, in fact), prevention, rehabilitation, and restoration do not get a quality standard of their own. And even though the government talks constantly about supporting independence, the Royal Commission recommendations wanted more equitable health for all, and all the Community Aged Care guides open with statements of health, wellness, reablement, and independence, exercise is not mandated in aged care (RCAQS). This frustrates me to the point of anger.

Internationally, multiple Position Stands advocate that older adults should be involved in twice weekly resistance training or weight-bearing exercise to maintain muscle function and bone integrity and should achieve 150 minutes of moderate or greater aerobic activity for heart and lung health (Chodzko-Zajko et al., 2009). The WHO (World Health Organization) states, 'Regular physical activity is proven to help prevent and manage noncommunicable diseases such as heart disease, stroke, diabetes, and several cancers. It also helps prevent hypertension, maintain healthy body weight, and can improve mental health, quality of life, and well-being' (WHO, 2000). The UK National Health Service takes this a step further adding that it can 'lower your risk of early death by up to 30%' (ULNHS). However, participation numbers continue to be low. Australian Institute of Health & Welfare (AIHW) data shows us that 70% of Australians over 65 years are insufficiently active (AIHW). What is concerning is that this data is self-reported, based on whether individual's feel that they are adhering to the published standards and guidelines. Many are doing their half hour walk per day (and, by many, I mean less than a third!) but the intensity of this walk is often too low to offer the promised benefits. The guidelines specifically state 'moderate or greater intensity'. But, using the 'talk test' which says a moderate intensity means taking a breath every three or four words, we can conclude the gaggle of older morning walkers who chat freely for their half hour saunter are not meeting this guideline. As a result of the population-wide phenomenon 'exercising-under-the-recommended-intensity', the WHO has suggested that the guidance for 150 minutes of aerobic exercise increases to 300 minutes per week.

At this point, the reader may have realised that we have not mentioned how many are doing the recommended resistance training? Well, I hope so anyway. You can fix everything with resistance training! This holds the most truth for an older adult who is having trouble getting off the toilet. Don't ever get dissuaded by the image of a muscle-on-muscle bodybuilder; resistance training is essential to build bone and muscle, improve balance, and combat functional disability for older adults! Yes, they will get bigger muscles but if they end up looking like Arnie, it would be a miracle or because they are abusing their steroidal medication. So back to how many are involved. Well, these figures are a little harder to find, with most surveys just concerned with physical activity, but Fragala et al. (2019) in their comprehensive review of resistance training reported that this number is as low as 8.7% of older Americans! If this is right and reflects engagement internationally, 91% of all older adults are missing the opportunity to stay strong, independent, and well. No wonder health systems are in such states of disrepair.

Why people are not involved is complex. I do not believe that it is for a lack of wanting to be well, but more due to low health literacy, poor government messaging, and lack of access/exposure to positive health-bringing options. Complicating things further is the common belief that disease and disability are part of the ageing process, with consumers and the medical specialist choosing medication and medical intervention treatment pathways over exercise and healthy lifestyles (like good nutritional intake and relaxation strategies). This has implications at the health system level where treatment is prioritised over prevention, and hospitals are rewarded for discharge and not whether the individual does not return to hospital. Post-hospital discharge programmes, such as the Transition Care Program, aim to get clients back to the level of independence that they had before hospitalisation, *not to a higher-level of function and/or into an ongoing program where the client can maintain or advance their gains.* Add to this that ageing is associated with increasing sedentary behaviour, which is shown to have deleterious health impacts (Biswas et al., 2015). Without maintenance programmes, clients can go backwards rapidly, often returning to hospital within six months. A recent independent analysis in the United States of participants in the nation's SilverSneakers' programmes, a Medicare supported National Initiative, identified 7.3 and 7.6 fewer hospital and emergency department presentations, respectively, among participants per year. The SilverSneaker group had lower hospitalisation risk, but if hospitalised they had shorter inpatient stays which equated to a 20% reduction in health expenditure (Crossman, 2018). With health systems and budgets under stress, and hospital beds in short supply, a system that is not listening to the overwhelming evidence and is content with treatment over prevention is broken!

Simply getting people to move more is an effective alternative to being discharged to do nothing. When this can be coupled with something that the individual enjoys, this request to move more can grow into regular dynamic and challenging activity participation, which people adhere to long term. Under this model, health improvements can be realised fully. Add to this a partnership with a specialist allied health clinician, experienced in prescribing exercise for older adults, and the individual will take steps towards improved and prolonged health. If this clinician then collaborates with a team of complementing clinicians, the individual can quickly be on and stay on their best health pathway. Using my service as the example, no matter how you come to us, you go through the same pathway. This starts with an initial comprehensive health assessment battery with the outcome brought to a case conferencing exercise. At case conferencing, we have eight different allied health professions represented, with the primary aim of ensuring that we have considered all the client needs. We ask: If they come for physio with a lower back issue, do they need podiatry due to a poor biomechanical foot strike that is impacting their pelvic alignment? Or, due to fluctuating undue fatigue, do they need to see our dietician? This interprofessional engagement in the restorative care plan guarantees holistic improvements. The journey always culminates in ongoing exercise participation. Depending on the starting point, it might take some older clients six weeks, and others six months, but the graduation to a twice weekly self-motivated group class exercise is a positive event worth celebrating. However, when compared to the significant body of evidence for the value of exercise (Longobucco et al., 2022; Matsuzawa et al., 2016; Yoo et al., 2022), the importance of an interdisciplinary approach to an individual's well-being is under studied. Sure, hospitals and other health system services claim to take the multidisciplinary approach. But, in my experience, it is GP/specialist-driven, clinician to clinician referral is poor, discharge planning is equally poor, and the client gets a backseat to the entire process. In our service approach, when allied health clinicians and therapists draw from their specific knowledge and work together to get a client onto their best health pathway, and do this in partnership with the client, the client can only benefit.

The health system may mention the value of exercise but do so in a cursory way, and clients are often not being referred to evidence-based health pathways. Word of mouth is the primary mode drawing older adults towards better health options. Nationally, healthy ageing has varying advocates ranging from individual GP to Primary Health Networks, but of Australia's 3,000 aged care providers, fewer than 10% see healthy ageing as a business priority (Craike et al., 2018). While a recognition of reablement has been slowly growing over the past five years, both the government and most aged care providers continue to prescribe domestic assistance and personal care as pathways to independence. For those providers who have introduced reablement programmes, many are prescribing to services with little to no adherence to evidence and/or leading practice initiatives, and even fewer have demonstrated their obligation to improve their client's health and well-being. How the evidence can be overlooked is astounding! Take residential aged care: one of the studies I was lucky enough to be part of was Jennie Hewitt et al. (2018a, 2018b) Sunbeam trial which compared resistance training and balance exercise to usual care among residential aged care adults. Following a two-phase intervention, exercise participants experienced a 55% reduced falls incidents rate. One of the largest reductions reported in this setting to date. In addition, the programme returned an incremental cost-effectiveness ratio of $22 per avoided fall per person, representing the cost-benefit of delivery over usual care. With 245,000 people in residential aged care, and a 50% falls rate and 40% multiple falls rate, the Sunbeam programme suggests significant savings. Yet, sector reform has just embedded the new Australian Nation Aged Care Classification (AN-ACC) tool into the residential aged care setting with no dedicated time for allied health service provision to deliver such valuable initiatives. I ask: Short-sighted?

A big piece in this puzzle is the individual's health literacy. One in ten clinical specialists have sufficient knowledge to successfully navigate the health system (Gilbert et al., 2013). In short, this implies that they have insufficient knowledge to guide clients towards their best health and/or aged care system options. If this number is low among clinical specialists, is it no wonder individuals believe that their only options are 'see their GP or specialist' or 'go to hospital'. Complicating things further is that the health and/or aged care system is not easy to navigate, even to the most experienced users. General Practitioners (GPs) do not know about aged care, aged care providers don't understand Medicare, private health insurers don't understand residential aged care, hospitals aren't aware of options to support meaningful discharge, and consumers aren't aware of 75% (or greater) of the available options. There is growing health literacy evidence about the clinician and client working better to understand a diagnosis, but little work trying to connect people to services that match their needs. Whether this is because we, as clinicians, are too busy to refer outside our services, or too proud to say that we do not know or that someone else could help, it doesn't matter. Poor health literacy is estimated to be costing health systems hundreds of millions of dollars each year. We also know that when people have better health literacy, they form better partnerships with their clinical team and have better health outcomes. However, while students engage in client-facing placements and/or role-playing in their undergraduate studies, the value of the client–clinician relationship is not prioritised to budding health professionals during this period as a primary learning mode. Knowledge is essential, but knowledge that supports clients through a partnership approach has better outcomes than prescription alone. The knowledge to navigate the health system requires significant exposure and experience. How we facilitate change in this space I am not certain, but it surely warrants discussion. Until we start this discussion, poor health literacy will continue to be a primary driver in the escalating national health system burden.

A starting point to better health literacy might be health behaviours education. Getting enough exercise, having a good diet and sleep hygiene, not smoking, and not drinking too much are proven markers for better health in later life. This is a conversation that all health professionals can have with their clients. Educating clients on these simple things is a step in the right direction. And, God forbid, they don't just mention exercise or diet, but make a referral to an exercise physiologist or dietician, respectively. When this becomes regular practice, then positive change is around the corner. Understanding health options and opportunities is a big-picture issue that must sit with the educators, state, and federal health systems if they want to see change. Governments have periodically marketed the deleterious impacts of smoking or a larger waist size, and who can forget the Grim Reaper and HIV advertisements, but they have rarely offered education on improved health literacy and are yet to get behind prescribed exercise adherence and environmental design to support participation in exercise as a primary counter to functional decline in later life. Health system policy, TV campaigns, and school curriculum modification all are solid starting points. As an extension, it would be amazing if hospitals embedded health educators to communicate with clients at their bedsides, married this to some simple-to-use resources and options for exercise engagement post discharge. This must be supported by positive discharge planning that leads to ongoing participation. If the hospital and aged care sector worked together, and aged care providers prioritised the client's health over the cleanliness of their floors. And the GP said, 'Let's try a walk around the block and add some chair stands before we try a new medication'. This would be amazing! I dream of a new world.

As I approach the end of my well-informed rant derived from a lifetime of dedicated work in healthy ageing and exercise participation, I reiterate: the evidence for exercise as a pathway to health and wellness for older adults is resounding. It is more than a walk around the block, it must incorporate the regular utilisation of a gym, environmentally designed with the older adult in mind, and weight-bearing exercise. Without this component, muscle and bone health atrophies, and people become frail. Complement resistance exercise with regular moderate-intensity aerobic exercise that gets the individual puffing and panting, their blood pumping, and heart and lungs working, and individuals can be on a course to better health and prolonged independence. However, this message is not told well. It must be shouted from the treetops. As we rattle along on the aged care reform roller coaster, influenced by a Royal Commission that highlighted the need for better health options, we need to ask, 'How can we help our clients?' Irrespective of whether the individual lives in residential aged care or the community, we all need to promote better pathways to health and well-being for older people (and start throughout the ageing life cycle, health literacy at all ages begets health literacy throughout life). Governments and state health systems need to listen to evidence, incorporate this into policy, educate their constituents, and hold providers responsible in their duty of care to clients. Providers need to offer evidence-based health options, take responsibility for clients' functional decline, and support their access to improved options to achieve better health. While older individuals need to get involved, their challenges may be the greatest due to addressing historic beliefs and expectations. They must achieve their health goals with the right support that motivates ongoing participation. Healthy ageing is complex in many ways but, like all things, we must start somewhere, and we must anchor efforts to evidence.

11.1 REFLECTIONS

The value of exercise as a modification of later life health and well-being first became apparent to me in my undergraduate. As a mature aged university entrant, I could easily recognise the resilience I had lost to bounce back after a weekend of drinking, social and competitive sport, when compared to my school colleagues. To regain this resilience, I focus on my own health, and found myself applying the question of the impact of exercise on ageing well to all my university assignments. It amazed me how much evidence there was and confused me that this was overwhelmed by the resistance to participate across so many levels. It took me a while to stop shouting from my soapbox 'get involved' at everyone. Instead, I embraced the opportunity to slowly chip away, first as an academic and then as a professional.

And here I am 27 years later, still amazed that this is not a mandatory part of the education system, that it is not advertised on TV, and that the health system doesn't prescribe exercise as the priority over medication, but I believe that it is getting better. I supported a mate at a local triathlon the other day and the people coming across the finish line came in all shapes and sizes. Similarly, when I support my wife at her long-distance trail running events, you would never assume this is what 90% of the finishers do on their weekends if you walked past them in the street. I see the same in the H&W gym, a whole lot of people getting involved in something completely foreign to them, but that they know is good for them without guidance by their GP and without shouting from the treetops about their involvement. This gives me hope that the mountain of evidence is not going unnoticed. The tables are turning and one day exercises will be a paramount, mainstream, and well-recognised positive modifier of health in later life.

REFERENCES

Australian Institute of Health and Welfare (AIHW). www.aihw.gov.au/

Biswas, A., Oh, P. I., Faulkner, G. E., Bajaj, R. R., Silver, M. A., Mitchell, M. S., & Alter, D. A. (2015). Sedentary time and its association with risk for disease incidence, mortality, and hospitalization in adults: A systematic review and meta-analysis. *Annals of Internal Medicine*, *162*(2), 123–132. https://doi.org/10.7326/m14-1651

Chodzko-Zajko, W., Proctor, D., Fiatarone Singh, M., Minson, C., Nigg, C., Salem, G., & Skinner, J. (2009). American college of sports medicine position stand: Exercise and physical activity for older adults. *Medicine & Science in Sports & Exercise*, *41*(7), 1510–1530. https://doi.org/10.1249/MSS.0b013e3181a0c95c

Craike, M., Britt, H., Parker, A., & Harrison, C. (2018). General practitioner referrals to exercise physiologists during routine practice: A prospective study. *Journal of Science and Medicine in Sport*, *2*(4), 478–483. https://doi.org/10.1016/j.jsams.2018.10.005

Crossman, A. (2018). Healthcare cost savings over a one-year period for silver sneakers group exercise participants. *Health Behavior and Policy Review*, *5*(1), 40–46. https://doi.org/10.14485/HBPR.5.1.4

Fragala, M., Cadore, E., Dorgo, S., Izquierdo, M., Kraemer, W., Peterson, M., & Ryan, E. (2019). Resistance training for older adults: Position statement from the national strength and conditioning association. *Journal of Strength and Conditioning Research*, *33*(8), 2019–2052. http://doi.org/10.1519/JSC.0000000000003230

Gilbert, A., Roughead, L., McDermott, R., Ryan, P., Esterman, A., Shakib, S., Luszcz, M., Vitry, A., Caughey, G., Preiss, K., Ramsay, E., Clark, A., & Zhang, Y. (2013). *Multiple chronic conditions in older people. Implications for health policy planning, practitioners and patients*. www.unisa.edu.au/siteassets/episerver-6-files/global/health/sansom/documents/qumprc/multiple-chronic-health-conditions.pdf

Hewitt, J., Goodall, S., Clemson, L., Henwood, T., & Refshauge, K. (2018a). Progressive resistance and balance training for falls prevention in long-term residential aged care: A cluster randomized trial of the Sunbeam program. *Journal of the American Directors Association*, *19*(4), 361–369. https://doi.org/10.14485/HBPR.5.1.4

Hewitt, J., Saing, S., Goodall, S., Henwood, T., Clemson, L., & Refshauge, K. (2018b, March). An economic evaluation of the SUNBEAM programme: A falls-prevention randomized controlled trial in residential aged care. *Clinical Rehabilitation*, *33*(3), 524–534. https://doi.org/10.1177/0269215518808051

Longobucco, Y., Masini, A., Marini, S., Barone, G., Fimognari, C., Bragonzoni, L., Dallolio, L., & Maffei, F. (2022). Exercise and oxidative stress biomarkers among adult with cancer: A systematic review. *Oxidative Medicine and Cellular Longevity*, *2022*, 2097318. https://doi.org/10.1155/2022/2097318

Matsuzawa, R., Hoshi, K., Yoneki, K., & Matsunaga, A. (2016). Evaluating the effectiveness of exercise training on elderly patients who require haemodialysis: Study protocol for a systematic review and meta-analysis. *BMJ Open*, *6*(5), e010990. https://doi.org/10.1136/bmjopen-2015-010990

National Library of Medicine: PubMed.gov. Retrieved January 20, 2023, from https://pubmed.ncbi.nlm.nih.gov/

Royal Commission into Aged Care Quality and Safety (RACQS). https://agedcare.royalcommission.gov.au/publications/final-report

Sherrington, C., Fairhall, N. J., Wallbank, G. K., Tiedemann, A., Michaleff, Z. A., Howard, K., Clemson, L., Hopewell, S., & Lamb, S. E. (2019). Exercise for preventing falls in older people living in the community. *The Cochrane Database of Systematic Reviews*, *1*(1), CD012424. https://doi.org/10.1002/14651858.CD012424.pub2

Tuckett, A., & Henwood, T. (2014). The impact of five lifestyle factors on nurses' and midwives' health: The Australian and New Zealand nurses' and midwives' e-cohort study. *International Journal of Health Promotion and Education*, *53*(3), 156–168. https://doi.org/10.1080/14635240.2014.978949

United Kingdom National Health Service (2021, 24 September). *Physical activity guidelines for older adults*. www.nhs.uk/live-well/exercise/exercise-guidelines/physical-activity-guidelines-older-adults/

World Health Organization (WHO) (2000). *The implications for training of the life course approach to health. (No. WHO/NMH/HPS/00.2)*. https://apps.who.int/iris/bitstream/handle/10665/69400/WHO_NMH_HPS_00.2_eng.pdf

Yoo, J., Ruppar, T., Wilbur, J., Miller, A., & Westrick, J. C. (2022). Effects of home-based exercise on frailty in patients with end-stage renal disease: Systematic review. *Biological Research for Nursing*, *24*(1), 48–63. https://doi.org/10.1177/10998004211033031

12 Returning to Work
The Impact of Healthcare on Workers Recovering from Workplace Injury

Nicholas Driver

12.1 MY BACKGROUND

My professional experience has exposed me to a vast range of people, professions, and risks. I have therefore seen the personal impact of physical and psychological injuries experienced by workers when things go wrong. For much of that time, my work has caused me to wonder about the interface between healthcare and the greater occupational context in which I work. I began to ask myself, 'What is unique about the relationship between a worker (as a patient) and their treating healthcare professionals? Is it different to other healthcare provider–recipient relationships?'

The Rehablitation and Return to Work Coordinator (RRTWC) acts as a conduit between the injured worker, their employer, their treating medical practitioners, and the workers' compensation insurer. Within Queensland, most employers are covered by a state-owned insurer, while other employers are self-insured through a variety of schemes. In other Australian states, workers' compensation insurance is provided by independent third parties, which can introduce intricacies to the Return to Work (RTW) process.

As a Work Health Safety (WHS) generalist, my work is divided among preventative hazard management, compliance work, incident investigation, and worker rehabilitation and return to work. This is made more complex by working for a large organisation with a diverse workforce involved in a wide range of work tasks and activities. Typically, the extent to which the employer provides a dedicated RRTWC is usually determined by the size of the workforce and the likelihood of multiple cases being simultaneously managed. It is not unusual for this role to be combined with other safety-related duties and responsibilities, as is the case in my role. I manage an average of six injured workers as part of my broader WHS generalist role. Though, at times, I have been responsible for the individual management of up to 14 concurrent claims for a wide range of physical and psychological injuries and illnesses. Because I am not a full-time rehabilitation specialist, when workers' compensation claims become numerous and complicated, they demand much more of my time and become a higher priority than my other activities. This prioritisation of the RTW process means that I am less accessible to my internal clients who often require timely risk management advice, and immediate and reactive responses to workplace incidents. The significant demands on my time to prioritise RTW can take an emotional toll on

DOI: 10.1201/9781032711195-14

me in my RRTWC role. This is especially true when I am managing psychological injury claims. While I focus my attention on holding true to a firm, objective, and fair claim management process, I am exposed to perceived occurrences of organisational misbehaviour or procedural injustice. In turn, these can add complexities to my general work duties or problematise the relationships I have with colleagues.

State regulators, insurers, and RRTWCs have applied significant effort to educate workers and their managers about the need for effective RTW processes (Safe Work Australia [SWA], 2019), and to remove some (but not yet all) of the stigma around work injuries. I see the next big frontier to be in the improvement of education and work design elements for medical professionals who deal with injured workers. Specifically, the average GP consultation lasts only a few minutes. For a workers' compensation case, much of that time is required to correctly complete the Work Capacity Certificate (WCC). This results in little time being available for actual diagnosis and treatment, and the WCC is regularly incorrectly filled in. Poor diagnosis, limited treatment, and incorrect completion of formal documents all have distinct negative consequences for a worker's return to work.

Considering my experiences through the lens of work design, I outlined my view of the successes and failings of a system that is, at the heart, intended to help people to recover from injury or illness. It is from this perspective that I offer a small number of claim case studies to demonstrate some of the concepts and struggles faced in the RRTWC context.[1] I hope that the stories shared will provide an insight into patient care and thereby contribute to the overall improvement of the patient–doctor relationship. Each case study highlights my experience of specific barriers to the achievement of my role's primary goal: return to productive work. While my work is outside the healthcare industry, I must manage the results of a worker's interactions with healthcare services. My suggestions could lead to process improvements and positive adjustments to work design principles for healthcare practitioners and specialists when they work with patients who have sustained work-related injuries. Further, these process improvements could also assist in meeting the targets of the National Return to Work strategy 2020–2030 (SWA, 2019).

12.2 WORKER ONE: COMPLEX TREATMENT NEEDS

Bill worked in an office-based role and was responsible for a small specialist team. On the day of injury, Bill was car-pooling with two other staff to attend a secondary work location a short distance from their main office. Bill and his colleagues were involved in a car accident while performing their work duties. Though investigations indicated that they were not at fault, the incident still resulted in an immediate and significant increase in the organisation's injury case management workload because all workers were injured. While Bill's injury was initially treated as minor, he soon presented with numerous difficulties while performing his doctor-prescribed suitable duties and the range of treatment needs became increasingly complex. Bill's physical injury seemed minor at first, meaning that he was able to perform many duties at a

[1] All individuals have been anonymised to protect their identity and privacy.

diminished capacity. However, Bill's progress towards returning to pre-injury capacity was delayed by limited response to therapy and side effects from the prescribed medication. This resulted in the development of secondary psychological concerns and pain management needs.

In my experience, Bill's claim was the first to reach double digits in the number of treating medical practitioners, with multiple practitioners in various fields consulted. This included general practice, physiotherapy, occupational therapy, psychology and psychiatry, orthopaedics, pain management, and neurology specialists. Other work-related issues presented complications in his injury management. For instance, there were allegations of misconduct and harassment of other staff, which required a fine balance when speaking with Bill's manager and the Human Resources team about his return to work. It was important that only relevant injury-related information was provided to other parts of the business. Overall, Bill's treatment lasted more than 18 months from the date of injury, which is significantly longer than was expected for this type of injury. Delays in Bill's recovery were disregarded as it was accepted that his injury would take time to heal. Over the duration of Bill's claim, he became increasingly negative about his treatment and eventually about his chances of a full recovery. This mindset was further impacted by obvious interpersonal issues occurring between Bill and his senior leader and peers, as well as the workers who reported directly to him. Bill's overall mood was potentially also affected by his positioning in an open-plan office area in which he had limited privacy, resulting in his injury being on display to other people based in the office or for those walking past. Given that the physical workplace can have significant impacts on the health and safety of individual works and their clients (Golembiewski, 2012), it is recommended that the immediate work environment must be considered when crafting suitable duties for office-based workers in future.

One of Bill's treating practitioners had a specific professional interest in Complex Regional Pain Syndrome (CRPS). Unsurprisingly, Bill was diagnosed with CRPS, and this formed the basis for much of his treatment. This diagnosis was later disproven by a surgical specialist who identified a previously undiagnosed physical element of the injury that clearly explained the pain and allowed for a meaningful change in the focus of Bill's treatment. However, cognitive biases are often associated with medical decisions (Saposnik et al., 2016), and this type of misdirection is not unusual and can be costly emotionally, through resources allocation, or to health outcomes. The health sector shares the phenomenon of vulnerability to human error with conventional industrial sectors (Zeltser & Nash, 2019).

Bill's case teaches us that it is important for treating practitioners to communicate effectively to ensure the timely and accurate exchange of information. This is not simply a case of Doctor A sending a report to Doctor B, or an expectation that the RRTWC or the insurer will collate records for a file. Rather, I suggest that patients such as Bill required a specific case manager who can take a lead role in the identification and implementation of their diagnosis, care, and treatment. Fundamentally, there is an opportunity for growth and further study as to the benefits of having a Primary Care Physician (PCP) (or someone like this with case management capability), not simply the patient's treating General Practitioner, taking the lead role in managing complex cases (Zeltser & Nash, 2019). Had Bill been under the direct care of a more astute practitioner earlier in his claim and had I – the RRTWC – had greater consideration of the salutogenic model (Antonovsky, 1979), I expect that Bill's perceived erroneous medical and professional treatment could have been altered earlier and a more targeted intervention could have aided Bill in a quicker recovery.

12.3 WORKER TWO: COMPLEX EMOTIONAL NEEDS

Jeffrey had several work injuries over a lengthy career of physical labour and operation of powered equipment. Jeffrey's career involving manual work exposed him to numerous hazards, many of which were compounded by social and lifestyle factors, and at least one significant injury not related to work. In some cases, new incidents resulted in the aggravation of a previous injury. During his return to work, and in independent investigations, Jeffrey was identified as having a strong external locus of control and a tendency to shirk responsibility. This resulted in a perception that his injuries and general treatment were a result of the wrongdoing of other people. Jeffrey stated that he was repeatedly treated poorly because he identified as a social minority. He expressed significant levels of distrust and blame towards many people involved in his treatment and return-to-work plan. His behaviour ranged from passive avoidance to aggressive outbursts regarding his perceived vilification, threats of self-harm, and admissions of substance misuse. He also complained that the insurer or medical professions were deliberately taking too long to process treatment interventions because they wanted him to suffer. Over time, it became progressively harder to manage Jeffrey's expectations in the workplace because each difficult interaction reinforced his negative expectations of others at work. His situation became increasingly complex when his lack of health literacy caused a misunderstanding about the relationship between work injuries, age-related degeneration, and lifestyle factors. On many occasions, Jeffrey found himself at odds with all support providers. Most importantly, various medical practitioners identified underlying medical issues relating to his work injuries. While medical advice suggested that significant surgical intervention was required for Jeffrey, it was made clear on multiple occasions by each treating practitioner that this intervention was not required as a direct result of the work injury. Whether by choice, a lack of understanding, or because of his collection of apparent injustices, Jeffrey had significant difficulty in reconciling the difference between his perception and the medical explanations for the surgery that he required.

Workers like Jeffrey present a most peculiar challenge. On the one hand, we have the most noble goal of returning the injured worker to their pre-injury role with as few permanent restrictions as possible. On the other hand, there is an increasing likelihood that their continued presence in the workplace is likely to cause negative outcomes for the health, safety, and well-being of both them and other workers. Many colleagues perceived Jeffrey's behaviour as an attempt to 'rort the system'. Further, his personal traits, mistrust of the organisation, and apparent refusal to accept responsibility for his health all conspired against him and any chance of future success. As described by Wall et al. (2009), his overall beliefs, external locus of control, and adversarial perspective contributed to negative rehabilitation outcomes. He was at risk of further injury while continuing to demonstrate aggressive and sometimes violent outbursts that placed other employees at risk of harm. Jeffrey's claim demonstrates the balance that must be maintained between the theoretical and the practical best outcome for the worker involved. A pragmatic approach requires a realistic consideration of long-term outcomes and, again, requires the involvement of a greater range of specialists to achieve a positive conclusion.

In Jeffrey's case, independent medical specialists were consulted, and their advice provided staunch support for his departure from the workplace to avoid future reinjury or aggravation. While this outcome is often seen as a failure, it was deemed to be in Jeffrey's best interests. I suggest that additional support, especially regarding health literacy earlier in his career, may have assisted Jeffrey to better understand his health status and provide a better Return to Work outcome. van Duijn et al. (2004) posit that the injured worker's education is one of those most common significant barriers to successful recovery and return to work. Further, an independent PCP and use of an independent, trusted support person not linked to the employer or insurer (like a case manager) may have assisted in increasing Jeffrey's health literacy and assisted in achieving a better outcome for him, regardless of the ongoing relationship with his employer.

12.4 WORKER THREE: COMPLEX MEDICAL RESTRICTIONS

Frank is the youngest of the workers described in these cases, and I suggest that his age played a critical role in many of the work injuries he sustained during his brief period of employment. Frank experienced physical injuries because of hazardous manual tasks, workplace interactions, and his work in hot and humid conditions without acclimatisation, training, or support to follow good work practice. Indeed, Frank provides a very real example of how different workers receive various levels of workplace support, depending on how well they are liked or respected by their superiors and peers. While each of his injuries were of a minor nature, concern was raised because of their number and frequency.

Frank's work injuries were initially regarded by managers as consequences of his poor attitude and work ethic. However, on closer analysis, there could be a vastly different explanation for his poor safety record. All reports suggested that Frank would never be 'Employee of the Month' – if such an award existed – and this common perception of him by work colleagues impacted the way that his supervisor and manager treated him relative to his work performance and his eventual work-related injuries. In fact, Frank's accident-prone nature was very much related to a simple lack of discernment as well as to supervisory and instructional failures that are vital for such a young and head-strong worker. Further, a late diagnosis of other non-work-related medical concerns may have explained some of his early injuries, which were fundamentally a matter of inattention or distractedness while undertaking basic tasks.

Conversations with Frank about his medical concerns highlighted a lack of general health literacy as he demonstrated extraordinarily little understanding of the seriousness of his injuries or the consequences of his failure to manage his health as directed by his medical practitioners on his ongoing employment. I believe that if he had a higher level of health literacy, he would have been more likely to seek medical assistance outside of work.

While healthcare professionals are acutely aware of the complex operation of the human body, they rarely have a detailed knowledge of the full range of work activities undertaken by individual worker's professions (Isernhagen, 2006). Such worker empowerment, as demonstrated in a salutogenic model focused on health (Antonovsky, 1979), could provide significant benefits for workers such as Frank. Had he developed a greater understanding and interest in his own conditions, Frank could have received treatment for the underlying medical physical and neurological conditions that seem to have contributed to his work-related injuries. This highlights the need for an expansive health system approach and support to improve health literacy amongst workers, especially those workers whose health issues have significant potential for aggravation in the workplace.

If the PCP is not afforded the time to assess patient health literacy and educate them accordingly, I suggest that someone else must be allocated that role. This is crucial to patient outcomes and return to work strategy.

12.5 WORKER FOUR: SIGNIFICANT TRAUMA

Brian is another young worker, but with trade qualifications and experience. Brian's case demonstrates how the overall success of rehabilitation is impacted by a worker's unique characteristics, care provision, and management of expectations. Brian sustained a significant limb injury while undertaking work. When Brian was transferred to a private hospital, his treatment was delayed because he held neither private health insurance or a Workers' Compensation claim number; he was admitted the day after his injury, but it typically takes longer for claims to be processed and a number allocated. More reminiscent of healthcare systems in other countries, Brian's experience suggests that payment was prioritised over the health of the individual. Subsequent infection in the surgical site also delayed Brian's return to work. Physical health issues were compounded by emotional and personal issues, including a perceived sense of entitlement and a deep resentment of his employer and the RTW process. Brian made repeated statements of blame towards the organisation and its representatives, despite significant efforts to improve his rehabilitation at a level much more than many involved in the case considered to be reasonably practicable.

It became clear after a brief period that Brian's resentment towards his employer was complicated by strained relationships outside of the workplace, including with his spouse and children. These were identified as barriers to the rehabilitation process and additional counselling support was provided, despite the absence of a clearly defined secondary psychological injury. However, as the trust relationship at work was never truly mended, Brian's return to his pre-injury role was short-lived and he exited the organisation of his own accord only a year later.

Brian's case identifies healthcare implications for in-patient hospital care, and the frequent need for psychological treatment later in the rehabilitation process. The initial lack of direct communication between Brian's surgical care providers and the employer (and, thus, insurer) delayed his treatment and played a significant role in the degradation of Brian's relationship with those parties. I propose that this could have been another opportunity for a PCP to take a lead role in a patient's treatment and recovery to ensure that the worker had access to relevant medical care with at least in-principal support from the workers' compensation insurer as soon as possible after the initial injury. This could have alleviated many of the issues faced by Brian and others involved in his claim and could have radically changed the claim's outcome.

12.6 WORKER FIVE: UNREASONABLE MANAGEMENT ACTION

Susanna was an ambitious worker with potential for job success through promotions. She was responsible for leading a small team and held other high-profile responsibilities across the organisation. She was well liked by her colleagues but experienced the occasional falling out. Workplace relations reached an all-time low when Susanna participated in a disciplinary process due to anonymous complaints received by her manager. Her heightened response to this process was unexpected by others involved. As a result, Susanna decompensated quickly and was certified as being unable to attend work after initial consultations with her medical practitioner.

Susanna presents the perfect case of an invisible injury, at least from a distance. Without visible physical injuries, many people treated Susanna as though no injury had occurred. In fact, her manager seemed oblivious to the fact that she could have any injury at all because she was seen acting 'normally' in public settings such as the local supermarket. Her psychological injuries of depression and anxiety resulted in a decreased ability for emotional and behavioural regulation. The severe mental distress caused by her treatment at work resulted in her becoming completely withdrawn from previous social and physical activities and subjected her to vulnerabilities, leading to substance misuse. Her psychological injury had a profound impact on her life in a regional centre, which was most noticeable because of Susanna's previously extensive involvement in the local community.

While Susanna received adequate care in the form of medication and psychological therapy from her medical practitioners, she presented a considerable opportunity for increased liaison between practitioners and the employer to determine a more pragmatic and effective response. A PCP may have been able to provide direct recommendations to Susanna and her employer to achieve a more realistic outcome that benefitted all parties. I believe that earlier termination of employment may have been a more realistic outcome that would have assisted Susanna to achieve healthcare success when combined with assistance to find alternative employment.

12.7 THE PERSONAL IMPACT AND STEPS FOR THE FUTURE

These are only some of the complex cases I have managed and they present very real impacts on me as a RRTWC. A sense of emotional and mental fatigue eventually sets in when the workers' compensation process fails to meet the noble goal. However, I feel that with subtle changes to work design, different outcomes would result in a greater feeling of success. When organisational leaders understand the true benefit of effective injury management, the function can be correctly resourced. However, this is a significant challenge for smaller organisations or organisations that do not prioritise the well-being of their workers.

Organisational change is mostly considered in terms of process or culture, but physically redesigning the workplace can also make significant changes. In my workplace, I have direct control over my immediate work area, but I also have a professional capacity to inform and advise organisational leaders in the design or redesign of broader workplace issues. Research exists and can be referenced to identify potential costs or benefits of contrasting philosophies, such as enclosed office areas versus open-plan, significant collaboration versus quiet spaces for personal focus (Marzban et al., 2020), or shared dining facilities compared with eating in isolation at our own desk. Interestingly, this personal impact is also affected by the nature and personality of the individual RRTWC. While some RRTWC's bring specialist qualifications in occupational therapy, physiotherapy, or other allied health fields, others, like me, take on a joint role or transition from a WHS generalist role to become a specialist in the workers' compensation space. All work injuries or illnesses result from workplace incidents or are compounded by work-related events. This can have a significant impact on the RRTWC's ability to separate the work-cause and the needs of the injured worker. When a worker's acts or omissions have resulted in injury, some in a joint RRTWC/WHS role find it difficult to shift their mindset to focus on rehabilitation in an ostensibly no-blame system. It is important that people working across these similar yet competing contexts can change their focus between the distinct roles.

> To look after my mental and physical well-being when dealing with difficult client workers and complex cases, I focus on the process. I advocate neither for the worker or the employer. Instead, I advocate for an equitable process.

This results in worker personality and obscure factors being less likely to impact me. It also helps me to avoid involvement in often parallel worker performance or disciplinary processes. Additionally, it helps to separate the worker's treatment and rehabilitation from any potential involvement or fault in relation to the incident that led to the injury in the first place.

I am proud of a particularly good success rate in terms of returning workers to their pre-injury position, but I cannot accept sole credit. Such success can only be achieved through positive interactions and the direct involvement of treating medical practitioners, organisational management, and the workers themselves. However, these other stakeholders must often work with limited specific training, time, and other resources. As such, there is a need to identify ways to improve the work design for medical professions – especially GPs – to give them more time for direct patient contact to allow greater diagnosis and treatment determinations. I suggest redesigning the role of the PCP – as an Injury Management Specialist – as a realistic improvement to work design in the workers' compensation interaction with the healthcare industry. This will assist with expectation management and improved health literacy for patients, as well as enriched patient management processes for healthcare professionals providing care to an injured worker.

The role of a PCP is to act as the first contact for a person with an undiagnosed health concern as well as providing continuing care of varied medical conditions, not limited by cause, organ system, or diagnosis. In Australia, these doctors are often referred to as General Practitioners. However, complex injuries present a workload that is unmanageable for most doctors to handle in addition to their normal case-loads. I suggest that the work design for general practitioners is inadequate in the context of workers' compensation and the resource availability of an occupational health physician is scant. In some cases – especially in regional or rural areas – the PCP role may be augmented by specialised injury management nurses or GPs with a specific professional interest in occupational health.

PCPs need a specific demeanour to balance the individual needs and desires of each patient. As with my personal involvement with many injured workers, the PCP needs to favour the treatment and rehabilitation process instead of simply meeting the desires of the patient. For too long, some doctors have taken the role of 'patient advocate' – sometimes referred to as 'Dr How Long?' – and they have given the patient whatever they wanted, instead of what they needed. Research confirms that an early return to work is most favourable in all circumstances (van Duijn et al., 2004; Isernhagen, 2006).

Worker empowerment and the implementation of the salutogenic model will lead an injured worker to focus more on their recovery and improvements in health. On many occasions, workers are unable to return to work in their pre-injury role. However, despite personal and organisational barriers, return to an alternative role is beneficial to injured workers (Beadles, 1996), allowing the worker to discover their own capacities to contribute in meaningful ways. With the right attitude and support, this can be achieved for workers with the most complex injuries, even if it means finding new ways of working outside of their substantive role, or with meaningful modifications and adjustments made to their substantive role and workplace.

A disturbing trend with some doctors, though, is their preference for several weeks of total incapacity as the first response to work injuries. This results in negative outcomes, especially for workers with poor health literacy. It also fails to correctly establish the patient's expectations about return to work in the short, medium, and long term. Patients' health literacy must be identified soon after commencing care. Where poor health literacy is identified, the PCP must be empowered and resourced to take a proactive approach, ensuring that the patient is provided with easily understood information about their injury or illness so that they can make an informed decision about their treatment. Fundamentally, this may require a change in funding model to allow for additional patient support outside of the medical consultation.

Salutogenesis as the focus and a *sense of coherence* is defined as

> a global orientation that expresses the extent to which one has a pervasive, enduring though dynamic feeling of confidence that one's internal and external environments are predictable and that there is a high probability that things will work out as well as can reasonably be expected.
>
> **(Antonovsky, 1979, p. 123)**

This helps people to utilise available resources to deal with tension and stress and helps to determine their movement on the health Ease/Disease continuum (Mittelmark et al., 2022).

This is a call to have a larger proportion of case management for workers' compensation must be handled by an increased provision of Occupational Physicians. These doctors – trained and funded specifically to deal with injuries and illnesses relating to occupational and industrial factors – are better suited to oversee the treatment of complex injuries sustained at work. This would, of course, require an increase in funding and resourcing to make Occupational Physicians more readily available, especially in regional areas, even if this is achieved through remote assessment.

The competence and compassion of medical practitioners are not being questioned here. Indeed, the cases discussed here were only able to achieve any level of success in RTW with the support of a wide range of highly skilled practitioners. I simply suggest that GPs require new perspectives, additional training, and an updated funding model to enable them to better facilitate effective return to work. As such, they can better partner with the injured worker and workplace coordinators as co-creators of an effective return to work process

12.8 SUMMARY

This chapter has profiled occupational rehabilitation vignettes to elaborate on the nexus of healthcare quality and resources in the context of work. The occupational health (and conventional health) care approaches impact on workplaces, workers, and their families, and the social constructs of employability and well-being. Concepts have been presented and recommendations have been made to aid effective work-related rehabilitation:

1. The immediate work environment must be considered when crafting suitable duties for office-based workers.
2. A pragmatic approach to occupational rehabilitation must include a realistic view of long-term outcomes with a range of specialist involvement.
3. A trusted person not linked to the employer or insurer, like a case manager, can help improve worker health literacy and help them better manage and respond to the process of occupational rehabilitation.
4. A worker's health literacy is critical to successful recovery and return to work; this must be a focus of intervention.
5. Specialised providers with an injury management focus are essential to affect positive outcomes; there must be an investment in these professionals and upskilling primary care physicians, nursing, and allied health teams accordingly.

REFERENCES

Antonovsky, A. (1979). *Health, stress, and coping*. Jossey-Bass.

Beadles, R. J. (1996). *Workers' compensation and rehabilitation outcomes: A study of injured forestry logging workers in Alabama* [Dissertation, Auburn University Publishing] (ProQuest Dissertations).

Golembiewski, J. (2012). Salutogenic design: The neural basis for health promoting environments. *World Health Design Scientific Review*, *5*(4), 62–68.

Isernhagen, S. J. (2006). Job matching and return to work: Occupational rehabilitation as the link. *Work*, *26*(3), 237–242.

Karanikas, N., Pazell, S., Wright, A., & Crawford, E. (2021). The what, why, and how of good work design: The perspective of the human factors and ergonomics society of Australia. In F. Rebelo (Ed.), *Advances in ergonomics in design: Proceedings of the AHFE 2021 virtual conference on ergonomics in design* (pp. 904–911). Springer.

Marzban, S., Candido, C., Mackey, M., Engelen, L., Zhang, F., & Tjondronegoro, D. (2020). A review of research in activity-based working over the last ten years: Lessons for the post-COVID workplace. *Journal of Facilities Management*. *21*(3), 313–333. https://doi.org/10.1108.jfm-08-2021-0081.

Mittelmark, M. B., Bauer, G. F., Vaandrager, L., Pelikan, J. M., Sagy, S., Eriksson, M., Lindström, B., & Meier Magistretti, C. (Eds.). (2022). *The Handbook of salutogenesis*. (2nd ed.). Springer.

Safe Work Australia (SWA). (2019). *National return to work strategy 2020–2030*. www.safeworkaustralia.gov.au/resources-and-publications/corporate-publications/national-return-work-strategy-2020-2030

Saposnik, G., Redelmeier, D., Ruff, C. C., & Tobler, P. N. (2016). Cognitive biases associated with medical decisions: A systematic review. *BMC Medical Informatics and Decision Making*, *16*(1), 138–138.

van Duijn, M., Miedema, H., Elders, L., & Burdorf, A. (2004). Barriers for early return-to-work of workers with musculoskeletal disorders according to occupational health physicians and human resource managers. *Journal of Occupational Rehabilitation*, *1491*, 31–41. https://doi.org/10.1023/B:JOOR.0000015009.00933.16

Wall, C., Morrissey, S., & Ogloff, J. (2009). The Workers' Compensation Experience: A qualitative exploration of workers' beliefs regarding the impact of the compensation system on their recovery and rehabilitation. *International Journal of Disability Management*, *4*(2), 19–26.

Zeltser, M. V., & Nash, D. B. (2019). Approaching the evidence basis for aviation-derived teamwork training in medicine. *American Journal of Medical Quality*, *34*(5), 455–464. https://doi.org/10.1177/1062860619873215

13 Biomechanics in Healthcare Design
Two Personal Journeys

Sahebeh Mirzaei Ezbarami and Mehrdad Hassani

13.1 AN EMERGING PHILOSOPHICAL ORIENTATION

During my bachelor's degree in medical engineering in Iran, I was passionate about applying engineering principles to solve medical problems, particularly in the field of biomechanics. The opportunity to combine my interests in engineering and healthcare led me to pursue these studies in medical engineering in which I was able to gain a solid foundation in topics such as anatomy, physiology, and medical device design. Throughout my studies, I was passionate about leveraging my knowledge of engineering to improve the lives of individuals with physical impairments while developing new and innovative medical technologies. To this day, this passion drives me to continue my education.

My desire to use engineering principles to better people's lives through revolutionary medical technology stems from my childhood when I vividly recall my grandparents' everyday challenges due to their decreasing health and limited mobility. Simple chores such as getting out of bed, walking, and even basic self-care duties were increasingly difficult for them. It was heartbreaking to witness their frustration and loss of freedom. To navigate their environment, they relied significantly on assistive devices such as canes, walkers, and wheelchairs, but these solutions frequently fell short of providing the needed degree of comfort and functionality.

Furthermore, the complexity and inconvenience of managing their medical illnesses, such as diabetes or heart disease, added to their stress. Witnessing their hardships had a profound influence on me and established my desire to use my engineering abilities to create revolutionary medical technology that could ease their suffering and improve their general well-being.

I pursued a master's degree in the field of medical engineering while I was still in Iran. Since the science of human factors and ergonomics in medical equipment was unfamiliar to me, I was curious and eager to delve into this arena. My ambition for designing effective solutions to enhance human biomechanics continued to grow. I found myself drawn to the field of rehabilitation engineering and ergonomics, which provided me with the opportunity to extend my skills and knowledge in this area. I have gained a deeper understanding of the complex interactions between functional movement, cognition, and physical rehabilitation because of my studies and practical experiences, and I am constantly motivated to find new and innovative ways to improve the lives of people with physical disabilities or impairments.

DOI: 10.1201/9781032711195-15

I have discovered the value of incorporating ergonomics into medical device design through my studies and practical experiences. Ergonomic designs in medical devices can improve device usability and comfort for both patients and healthcare workers, lowering the risk of injury or pain during use. In an Iranian hospital, for example, I witnessed a nurse assisting a patient in altering their position using a standard hospital bed. Many of the beds in the facility shared similar characteristics: they were heavy, difficult to manage, and lacked user-friendly features. I could see the effort on the nurse and her back strain as they attempted to lift the head of the bed. The patient, a kind-hearted Iranian, also voiced pain because the bed's stiff structure did not fit their body shape.

> Consider an alternative scenario in which medical equipment with an ergonomic design is used. In this scenario, the nurse might operate a user-friendly adjustable bed that was designed with the patient and the healthcare professional in mind. As the nurse adjusts the bed with ease, they notice that it reacts seamlessly to their directions, eliminating physical strain on their body. At the same time, the patient feels more at peace because the bed can mould to their body, offering ideal support and comfort.

Pursuing a master's degree in this profession has been a transforming experience that has broadened my technical knowledge and profoundly influenced my personal growth. This path altered my perspective on cultural influences and societal expectations, particularly as a woman in Iran. Cultural conventions and gender expectations frequently impede women's educational and professional potential. I was confronted with societal pressures and preconceptions that limit the roles and pathways open to women (like the many assumptions made by men and women that women were not capable in these fields or would be wise to reinvest their time in raising a family). As a female, I have experienced the pervasive presence of assumptions and biases regarding women's capabilities and priorities. These assumptions can be seen in many areas of life, such as education, career choices, and cultural expectations.

However, my enthusiasm for engineering and drive to break free from these restraints spurred me onward.

13.2 TRANSFERRING SCIENTIFIC KNOWLEDGE TO THE INDUSTRY

After graduating, I contemplated what would fulfil my desire to be of service to others. I realised that pursuing a doctorate degree, deepening my knowledge through extensive research, and sharing my expertise with others would satisfy this inner drive. I did not make the decision to seek a PhD degree lightly, especially given the added hurdles that women experience in Iran when pursuing higher education. Women in Iran may encounter limited access to educational resources, research opportunities, and funding. Sexual orientation prejudices and preconceptions persist in Iranian academic institutions, deterring women from obtaining their PhDs. These prejudices may emerge as extra performance expectations, receiving less credit for their accomplishments as their male counterparts, and receiving limited support from instructors or advisors.

My strong academic background, coupled with my enthusiasm for teaching and sharing knowledge, caught the attention of faculty members and professors within the university. It was difficult to gain acceptance as an educator with the various obstacles to overcome. There were times when people questioned my aptitude for the post merely because I was a young professional woman in Iran. Some questioned my capacity to bear the demands of teaching, reflecting stereotypes about women's academic talents. Furthermore, my opinions and contributions were sometimes discounted or rejected. Despite these challenges, I remained determined and resilient, demonstrating my knowledge and commitment to my field.

Fortunately, with the passage of time, my aptitude as an educator was acknowledged and I was given the chance to teach a variety of subjects within my field. Thus, I enrolled in a doctoral programme in Iran and simultaneously began teaching various subjects related to my field, including human factors, biomechanics, ergonomic design, and medical equipment at the university.

As a lecturer, I was intimately aware of the unique hurdles that women confront in Iran when pursuing education and jobs in professions traditionally dominated by men. Now, I strive to foster an open and supportive environment in which female students can overcome obstacles and thrive in their academic and professional goals. Serving as a mentor and role model for female students is one of the ways I address the issues that I encountered as a woman in Iran. I recognise the value of representation and the influence it can have on encouraging and empowering women to achieve their dreams. I share my experiences, problems, and accomplishments with female students navigating their educational careers through open and honest talks.

Furthermore, I actively seek collaboration possibilities with organisations and projects that promote women's education and empowerment. I hope to dismantle barriers and provide resources and support to Iranian women seeking education and careers in STEM subjects by participating in outreach programmes and community activities. I believe in the potential of collective action and collaboration to effect positive change and provide more fair opportunities for Iranian women. On the other hand, teaching provided me with an opportunity to stay up to date with the latest advancements in my field and gain new insights from my students' perspectives. I believe that the exchange of ideas and knowledge between students and instructors is a two-way process that benefits both parties.

My goal was to see the practical application of my research beyond academic publications. It was important for me to witness the impact of my years of study and research in the industry, and not simply be limited to writing and research on paper. As a result, I strove to bridge the gap between academia and industry by seeking opportunities to collaborate with companies and apply my research in real-world settings. Bridging the gap between academia and industry requires a proactive and strategic approach. Here are some steps and advice I would advise to students or individuals entering their careers to effectively bridge this gap:

- *Keep Up with Industry Trends:* Stay current on the newest breakthroughs, trends, and problems in your field of interest. This can be accomplished through continual learning, attendance at industry conferences, membership in professional networks, and reading industry periodicals. Understanding the industry's needs and priorities can help you focus your research and abilities accordingly.

- *Collaborate with Industry Partners:* Look for possibilities to work on research projects or internships with industry partners. Collaborative research with industry personnel provides vital insights into real-world problems and improves knowledge transfer between academia and industry. Seek out industry-sponsored research programmes, partnerships, or grants to help fund collaborative initiatives.
- *Effectively Convey Your Research:* While academic papers are important, it is critical to present your findings in a way that is accessible and relevant to industry personnel. Develop good presentation and communication abilities to successfully communicate the practical implications and implementations of your study. Highlight the potential effect and benefits that your work can offer to industry stakeholders.
- *Collaborate with Technology Transfer Offices:* Many colleges and universities have technology transfer offices that help academic research be commercialised. Engage with these offices to learn about patenting, licensing, and commercialising your research findings. They can guide you through the process of translating your research into practical applications that the industry would accept.
- *Be Open to Multidisciplinary Collaborations:* Industry problems frequently necessitate interdisciplinary solutions. Collaborate with experts from other fields, such as engineering, business, or design, to create holistic solutions that satisfy industry demands. Accepting interdisciplinary cooperation broadens your horizons and strengthens your capacity to bridge the gap between academia and industry.

13.3 FINDING A PERFECT FIT: ERGONOMICS AND MEDICAL DEVICE COMPANY IN IRAN

Having struggled to find a suitable company in the healthcare industry in Iran, I eventually landed a position in a medical and rehabilitation equipment company that shared my passion for ergonomics. Despite my skills and enthusiasm for this specialised field, I was disappointed to discover a lack of knowledge of the value proposition of human factors in medical device design. During my job search, I saw how some companies underestimated the need of incorporating human aspects principles into medical device development. Many of them overlooked the importance of ergonomic design and user-centred methods in enhancing patient outcomes and healthcare professional experiences. This lack of comprehension frequently resulted in limited work opportunities and a sense of frustration. It was frustrating to see the industry's reluctance to recognise the significance of end-user experience and the influence it can have on device usage, safety, and patient satisfaction. While I came across a few progressive companies who recognised the importance of human aspects, they were frequently outweighed by organisations that did not completely embrace these ideas. These difficulties, however, did not deter me. I persisted with my employment hunt, aiming to find a company that shared my vision and recognised the importance of human considerations in medical device design. Finally, I landed a position with a forward-thinking medical and rehabilitation equipment company that valued user-centred design and the influence it could have on enhancing patient care.

This experience stimulated my desire to advocate for the incorporation of human aspects in medical device development and educate others on the benefits. I realised that by increasing knowledge and promoting the significance of human aspects, we can drive industry-wide change and build a future in which healthcare technology prioritises user requirements and experiences.

In addition to my role as an equipment biomechanics expert, I also served as an ergonomics consultant for the team. The company I worked for had a dynamic team of young, motivated, and creative individuals who were driven to spot industry gaps and develop unique solutions to fill them. We focused on accessibility and ease of use, customisation, and adaptability, and in all situations, the team used an approach to design that included thorough user research, prototyping, and iterative testing. We actively engaged healthcare professionals, patients, and caregivers throughout the design process to gain views and feedback. This collaborative approach guaranteed that the solutions they generated were not only technically novel but also practical and fit with the unique demands of the end users. Working in such a setting was very exciting for me because my co-workers and I shared a common aim of enhancing the health of the Iranian people.[1] My reflections on these experiences include the following:

- *Challenges:* Due to gender biases and cultural expectations, Iranian women may encounter specific problems on the job. Despite improvements, there may still be a lack of women in leadership roles, and gender stereotypes may exist.
- *Work-Life Balance:* Family and community values are highly valued in Iran. Women frequently balance their career ambitions with home and social obligations. This can imply juggling domestic responsibilities while pursuing a profession, which may necessitate juggling different priorities.
- *Socialising Cultural Norms:* There are cultural standards on appropriate interactions between men and women in social settings. This can make it difficult for women to begin or sustain professional relationships since they must negotiate societal expectations while avoiding any appearance of impropriety.
- *Modesty and Hijab:* Iran has rigorous dress standards for women, including the wear of the hijab (outer garments), which might deter social interactions. Although women may feel more at ease following these principles, it can often create a barrier to casual contact and limit opportunities for informal networking.

It is critical to recognise that experiences in Iran may differ between individuals and among organisations. While there are some hurdles, Iranian women show tenacity and

[1] A professional woman's life in Iran can be a whirlwind of chances, problems, and cultural influences.

determination in pursuing their professional goals and contributing to their fields of competence. Iranian women's increasing presence and achievements in numerous fields reflect their continued contributions to the development and advancement of Iranian society.

13.4 REVOLUTIONISING MOBILITY: COLLABORATIVE INNOVATION IN HEALTHCARE FOR ENHANCED ACCESSIBILITY AND USER-FRIENDLY MEDICAL DEVICES

Together, we collaborated and produced innovative solutions to address the various challenges in the healthcare industry. We worked tirelessly to improve the quality of medical and rehabilitation equipment in Iran, and to make these tools accessible to as many people as possible.

One noteworthy case study that highlights our teamwork and creative problem-solving in the healthcare business was the development of a more accessible and user-friendly medical gadget for people with limited mobility. We realised that traditional mobility aids, such as walkers, frequently posed difficulties for people with mobility impairments, particularly those with diminished grip strength or trouble moving. We wanted to create a gadget that could increased support and manoeuvrability while being straightforward and simple to use.

Our team did comprehensive research, speaking with healthcare experts, patients, and carers to better understand their individual requirements and problem spots. We observed patients' challenges throughout mobility exercises and received input on existing equipment. Based on these insights, we developed a revolutionary design concept: improved mobility assistance with ergonomic handles, intelligent braking mechanisms, and configurable features to fit individual comfort and support demands. Lightweight materials and inventive engineering were used to ensure ease of usage and manoeuvrability. The final solution was the culmination of our collaborative effort, which combined engineering knowledge, user-centred design concepts, and stakeholder feedback. The mobility aid's revolutionary features and user-friendly design improved the mobility experience for those with limited mobility, increasing their independence and quality of life.

13.5 NAVIGATING UNIQUE HEALTHCARE NEEDS IN IRAN: ADDRESSING NON-COMMUNICABLE DISEASES, AGEING POPULATION, CULTURAL INFLUENCES, ACCESS, AND MENTAL HEALTH

People's healthcare demands in Iran, like those in any other country, are determined by a variety of factors such as population demographics, prevalent diseases, socioeconomic conditions, and cultural traditions. While it is crucial to highlight that healthcare demands differ between individuals and locations, there are some factors that are distinctive or noteworthy in the context of Iran:

- *Noncommunicable Diseases:* Noncommunicable diseases (NCDs) such as cardiovascular disease, diabetes, cancer, and respiratory problems are on the rise in Iran, as they are in many other nations. These disorders represent serious health risks and necessitate comprehensive healthcare approaches to prevention, diagnosis, and treatment.

- *Population Ageing:* With an ageing population, Iran is facing demographic changes. This demographic shift creates new healthcare needs in the areas of geriatric care, age-related diseases, and healthy ageing promotion. A priority is to provide proper healthcare services and support to the senior population.
- *Cultural Influences:* Cultural norms and practises in Iran might influence healthcare requirements and choices. Gender roles, family relationships, and religious beliefs all can have an impact on healthcare-seeking behaviour, treatment choices, and access to healthcare services. Recognising and managing these cultural factors is critical for good healthcare delivery.
- *Access to Healthcare:* In Iran, ensuring access to quality healthcare services for all sectors of the population, especially marginalised communities, and rural areas, is a persistent challenge. Improvements are being made to healthcare infrastructure, healthcare coverage, and healthcare delivery systems.
- *Mental Health:* As mental health concerns affect people of all ages and socioeconomic backgrounds, mental health awareness and services are gaining traction in Iran. Mental health promotion, prevention, and support services are becoming increasingly important.

To better fulfil the healthcare requirements of the people of Iran, we need to take a multidimensional strategy that includes needs assessment, user-centred design, adaptation of technology, collaboration with stakeholders, and continuous improvement.

Through these methods, we were able to tailor our equipment and services to address the specific challenges and requirements identified in the healthcare needs assessment. This approach allowed us to create solutions that were better suited to the local context, more effective in meeting the needs of healthcare providers and patients and aligned with the available resources and infrastructure.

In the end, our efforts paid off, as we were able to make a meaningful impact in the lives of countless individuals across the country by redesigning some medical appliances. For a period of approximately five years, I was employed in the industry. Throughout this time, I fulfilled the role of a biomechanics specialist, while simultaneously engaging in scientific research. I held the belief that continued education and research would prove instrumental in identifying and addressing industry gaps.

13.6 ACADEMIC/INDUSTRY RESEARCH HIGHLIGHTS

In collaboration with my friends and colleagues, I authored articles pertaining to occupational health, which were published in academic journals. One of them was 'Evaluation of Working Conditions, Work Postures, Musculoskeletal Disorders and Low Back Pain among Sugar Production Workers' (Hassani et al., 2021a). We conducted thorough assessments of the working conditions and body postures of employees involved in sugar production through on-site inspections and measurements. Additionally, we utilised structured questionnaires to gather comprehensive data on demographic information, work-related factors, and the occurrence of musculoskeletal issues. The results of the study showed that the workers were exposed to a high number of risk factors for musculoskeletal disorders and low back pain. These risk factors included repetitive movements, prolonged standing and sitting, awkward postures, and manual handling of heavy loads[2] (Hassani et al., 2021a).

Another one that I appreciate is the 'Prevalence of Musculoskeletal Disorders, Working Conditions, and Related Risk Factors in the Meat Processing Industry: Comparative Analysis of Iran–Poland' (Hassani et al., 2021b). The study encompassed the selection of a representative sample of meat processing workers from both Iran and Poland. Detailed information on the workers' demographics, job characteristics, and musculoskeletal symptoms was gathered using structured questionnaires. The objective was to compare the working conditions between a developed country and a developing country.

This study highlights the need for increased attention to ergonomic factors in the meat processing industry to improve the health and well-being of workers and reduce the economic burden of MSDs on both workers and employers. During my collaboration with researchers from Poland on the prevalence of musculoskeletal disorders in the meat processing industry, we discovered that differences in the devices used, as well as variations in workers' and researchers' mindsets, can lead to workplace problems.[3]

And the next one is 'Redesign and ergonomic assessment of a chest-support baby walker'. As we know, a chest-support baby walker is a device that helps infants learn to walk by providing support for their chest while they move their legs. However, traditional baby walkers have been associated with safety risks such as falls, injuries, and even fatalities. Therefore, a redesign and ergonomic assessment of a chest-support baby walker is important to ensure the safety and comfort of the infant (Golmohammadpour et al., 2023).

The first step in the redesign process was to conduct an ergonomic assessment of the current design. This involves evaluating the dimensions, materials, and functionality of the existing chest-support baby walker. The assessment will identify any ergonomic issues, such as discomfort or unsafe postures, and suggest potential improvements.

Based on the assessment, a new design can be developed that addresses the identified issues. For example, the new design may include adjustable straps and padding to ensure a comfortable fit for infants of different sizes. Additionally, the walker was made with lightweight and durable materials to make it easier to manoeuvre and transport.

The redesigned chest-support baby walker should also incorporate safety features to prevent accidents:

- *Staircase Falls:* One common accident is when a baby using a traditional walker falls down a flight of stairs. Traditional walkers with wheels can easily roll downstairs, putting the child at risk of serious injuries (Mohd Basar et al., 2021).

---

[2] I noticed that the manual cutting of sugar beets required the use of a sickle, which was not designed with ergonomic principles in mind.

[3] This highlighted the need for more comprehensive guidelines and protocols to ensure the safety and well-being of workers in this industry.

- *Tip-over Accidents*: Baby walkers with inadequate stability or uneven weight distribution can tip over, causing the child to fall or become trapped underneath. This can lead to injuries such as head trauma, fractures, or entrapment (Ducrocq et al., 2006).
- *Collision Hazards:* Uncontrolled movement and speed of traditional walkers can cause collisions with furniture, walls, or other objects, leading to injuries such as bumps, bruises, or cuts (Lueder & Rice, 2007).
- *Drowning Hazards:* If a baby walker falls into a swimming pool, bathtub, or other water sources, there is a risk of drowning if the child is not promptly rescued (McDonald et al., 2018).

Once the new design was developed, it was tested to ensure that it meets safety and ergonomic standards. This involved user testing with infants and their caregivers to evaluate the usability, comfort, and safety of the new design. The redesign of the chest-support baby walker was accomplished by a collaborative effort including a team of experts, which may have included engineers, designers, ergonomists, and child development healthcare professionals.

The development process involved various stages:

- *Conceptualisation:* The team brainstormed and created concepts for the new design, considering elements such as safety, ergonomics, ease of use, and developmental appropriateness. This stage comprises sketching, 3D modelling, or prototyping to visualise the design concept.
- *Iterative Design and Refinement:* Based on feedback from the team, experts, and future users, the basic design concept underwent numerous iterations and refinements. This iterative method aids in the identification and correction of any potential design flaws or enhancements.
- *Compliance with Safety Standards:* Throughout the development process, the design team worked to verify that the revised baby walker met all applicable safety standards and regulations. These requirements include recommendations for stability, materials used, locking systems, and other baby walker safety considerations.
- *Ergonomic Assessment:* Ergonomic assessment entails determining how well the revised baby walker matches the anatomical and physiological characteristics of infants, encourages good posture, and supports their natural movement. This evaluation may involve ergonomic concepts, anthropometric measurements, and consideration of age-appropriate design aspects.

Through these efforts, we were able to contribute to the ongoing discourse and share our findings with a broader audience.

13.7 IRAN TO AUSTRALIA: AN INTERCULTURAL JOURNEY OF LOVE, EDUCATION, AND PERSONAL GROWTH

When I made the decision to continue my education, I moved to the east of Iran to pursue my studies. It was during this time, while attending university or engaging in academic activities, that I had the opportunity to meet my future husband. Our paths crossed in this educational setting, and we began to interact and spend time together. As we got to know each other better, we discovered common interests, shared values, and a strong connection. Our friendship developed into a deeper relationship, and we realised that we wanted to build a life together. Recognising the potential of our bond, we decided to take the next step and formalise our commitment by getting married.

After our marriage, our journey led us to Australia, where my husband had been awarded a doctoral scholarship at an Australian university. This move provided an exciting opportunity for both of us to explore a new country and immerse ourselves in an international setting. For me, it was a particularly fascinating experience as I had always been eager to apply my knowledge and perspectives in a global context. I was enthusiastic about embracing the similarities and differences between Iran and other countries, and this journey allowed me to do that.

There are some reasons why I choose to come to Australia:

- *Education and Research Opportunities:* Australia is well-known for its excellent educational system and research organisations. Many overseas students opt to study in Australia because the country provides a diverse selection of academic programmes and scholarships.
- *Economic Opportunities:* Australia's economy is strong, with diversified industries that provide work opportunities and a positive business environment. Individuals seeking improved professional opportunities and financial security are drawn to the country's steady economy and quality living.
- *Natural Beauty and Lifestyle:* Australia is famous for its breathtaking scenery, which include magnificent beaches, rare fauna, and huge desert regions. The country provides a varied selection of outdoor activities as well as a laid-back lifestyle that many people find appealing.
- *Multicultural Society:* Australia is also a multicultural culture that values diversity and provides opportunities for people of diverse cultural origins to coexist and develop. This multiculturalism has the potential to create a vibrant and welcoming atmosphere for people of all nationalities and religions.

Now I think my experiences and knowledge in the field of medical devices will help me to adapt to the Australian culture, industry, and academic system. As someone who is passionate about medical equipment and biomechanics, I am excited about the potential opportunities and challenges that lie ahead. I believe that my experiences and knowledge in these fields will prove to be valuable assets in my pursuit of further education and career advancement in Australia.

Moreover, living in a new country has given me a chance to broaden my horizons and learn about different cultures. It has been both challenging[4] and rewarding to immerse myself in a new environment, but I am grateful for the experiences and friendships that I have gained along the way.

[4] Because of the considerable distance, I missed my family and friends, especially my mother.

My husband, who has a comparable academic background to mine, has also given his thoughts on improving medical device design by incorporating human factors and ergonomics principles. I invited him to express his thoughts on the possible advancements that future technologies could bring to the healthcare business.

13.8 MEHRDAD'S REFLECTIONS

13.8.1 Improving Medical Device Design with Human Factors and Ergonomics

Medical devices have revolutionised healthcare by improving patient outcomes (Mohd Basar et al., 2021), reducing recovery time (Ducrocq et al., 2006), and increasing efficiency (Lueder & Rice, 2007). However, poorly designed medical devices can lead to user errors, adverse events, and even patient harm. To mitigate these risks, it is crucial to incorporate human factors and ergonomics (HF/E) principles in the design of medical devices.

HF/E is the scientific discipline that focuses on optimising the interaction between humans and technology to enhance performance, safety, and user experience. In the context of medical devices, HF/E principles aim to design devices that are intuitive, easy to use, and promote user safety.[5]

One of the primary benefits of incorporating HF/E principles in the design of medical devices is improved usability. Medical devices are often complex and require specialised training to use effectively (McDonald et al., 2018). However, by designing devices with the user in mind, manufacturers can improve device usability, reduce training time, and increase user satisfaction. For example, a study on the usability of insulin pumps found that incorporating HF/E principles led to a reduction in user errors and improved user satisfaction (Mohammed-Nasir et al., 2023).

Furthermore, HF/E principles can help mitigate the risk of adverse events caused by medical device user errors (Minopoulos et al., 2023). Adverse events can occur when users make errors in operating the device or misinterpret device feedback. By designing devices with intuitive interfaces and clear feedback mechanisms, manufacturers can reduce the risk of user errors and improve device safety. For example, incorporating HF/E principles in the design of infusion pumps has been shown to reduce the risk of medication errors (Mohammed-Nasir et al., 2023).

In addition to improving usability and safety, incorporating HF/E principles in the design of medical devices can also improve device performance. This includes improving device accuracy, reducing device downtime, and enhancing overall device functionality.

In conclusion, manufacturers should consider the user's needs and capabilities when designing medical devices to ensure that they are easy to use, promote user safety, and optimise performance. This includes creating user-friendly user interfaces, offering transparent feedback systems, and reducing the cognitive effort of the gadget. Manufacturers can ensure that these gadgets continue to revolutionise healthcare while lowering the risk of user errors and adverse events by developing medical devices with HF/E principles in mind.

---

[5] This includes considering the device's physical design, cognitive workload, and user training.

13.8.2 The Potential of Emerging Technologies in Healthcare

Emerging technologies have the potential to transform healthcare.[6] In recent years, there has been a surge of interest in emerging technologies such as wearables, virtual and augmented reality, and robotics. I have had the chance to work with some of these technologies as a practitioner, and I am aware of both the advantages and the difficulties of doing so in the field of healthcare.

Wearable technology has gained significant attention in recent years for its potential to monitor health and wellness continuously (Cheng, 2003). Wearables such as smartwatches and fitness trackers can provide users with real-time feedback on their physical activity, heart rate, and sleep patterns. These devices can help users make informed decisions about their health and promote healthy behaviours. However, creating wearables that are cosy, user-friendly, and aesthetically beautiful can be difficult. To ensure that customers will wear the item frequently, manufacturers must consider elements including device size, weight, and materials.

Virtual and augmented reality (VR/AR) is another emerging technology that has the potential to transform healthcare. VR/AR can be used for medical training, patient education, and even pain management. For example, VR/AR can be used to simulate surgical procedures, allowing medical students to practice complex operations before performing them on patients (Webster & Haut, 2024; Ahmed & Ahmed, 2023). VR/AR can also be used to create immersive experiences that distract patients from pain and discomfort during medical procedures (Hall et al., 2020). Designing immersive, user-friendly, and efficient VR/AR experiences can be difficult, though. To ensure that the VR/AR experience is as good as it can be, manufacturers must consider elements like display resolution, tracking precision, and user interface design.

Robotics is another emerging technology that is transforming healthcare. Robots can be used for a variety of healthcare applications, such as surgery, rehabilitation, and patient care (Jin et al., 2017). Robotic movements can be accurate and reliable, lowering the possibility of human error and enhancing patient outcomes. Developing robots that are both safe and efficient can present a considerable challenge. Manufacturers must consider a variety of factors, including the size, shape, and mobility of the robot, to ensure that it can perform its designated tasks effectively and without posing a risk to human operators.

Designers can optimise user experience and outcomes while lowering the risk of user mistakes and unfavourable events by building these technologies with user capabilities and need in mind. These technologies have the potential to change the way healthcare is delivered while also improving patient outcomes.

13.8.3 Chief Work Design Strategist: An Integral Role in Incorporating Human Factors and Ergonomics in Medical Device Design

Medical devices are crucial in healthcare, but poorly designed devices can lead to user errors, adverse events, and even patient harm. Hence, principles of HF/E must

[6] Through strengthening overall healthcare delivery, lowering costs, and enhancing patient outcomes.

be incorporated into the design of medical equipment to reduce these dangers. The role of a Chief Work Design Strategist (CWDS) is increasingly critical in achieving this goal.

The CWDS can be responsible for ensuring that medical devices are designed with the user's needs and capabilities in mind. They work closely with manufacturers to optimise device usability, safety, and performance. One of the CWDS's primary responsibilities would be to incorporate HF/E principles into the design process. This entails creating user-friendly interfaces, offering clear feedback systems, and minimising the cognitive effort placed on the device.

The CWDS would play a crucial role in ensuring that medical devices are designed to accommodate a diverse range of users. This includes users with varying physical, cognitive, and sensory abilities. By designing devices that are inclusive and accessible, manufacturers can improve device usability, safety, and performance for all users.

The CWDS must also consider the environmental and contextual factors that may impact device performance. This includes creating gadgets that can resist a variety of environmental conditions, like changes in temperature and humidity. Additionally, the CWDS must consider the context in which the device will be used, such as in an emergency room or in a patient's home.

Making sure that medical devices are made to reduce the possibility of user mistake and unfavourable outcomes is another key duty of the CWDS. This involves developing devices with simple feedback systems and intuitive user interfaces. Manufacturers can lower the likelihood of user errors and enhance device safety by creating devices that are simple to operate and encourage user safety.

In summary, the appointment of a Chief Work Design Strategist position is crucial for integrating ergonomics and human factors into medical device design. Industries can enhance device usability, safety, and performance by considering user requirements and capabilities, environmental and contextual elements, and designing devices to reduce the risk of user mistake and adverse occurrences. Medical device innovation in healthcare depends critically on the good design to reduce the risk of user error and adverse results.

13.9 THE SUCCESSFUL CANDIDATE WILL

- The successful candidate for the CWDS position should possess a strong background in human factors and ergonomics, as well as experience in medical device design and development. They should have a deep understanding of the regulatory landscape and industry standards related to medical devices, as well as a keen awareness of user needs and capabilities.
- Additionally, the successful candidate should have strong leadership skills and be able to collaborate effectively with cross-functional teams, including engineers, designers, and regulatory affairs specialists. They should be able to develop and implement strategies that integrate human factors and ergonomics throughout the product development process, from concept development to post-market surveillance.

- Excellent communication skills are also essential, as the CWDS will need to effectively communicate the importance of human factors and ergonomics to stakeholders across the organisation. They should be able to present complex information in a clear and concise manner and be able to build consensus around design decisions that prioritise user needs and safety.
- The ideal applicant for this position will be passionate about enhancing patient outcomes and have a solid grasp of how ergonomics and human factors relate to one another.

13.10 SUMMARY: A SHARED VIEW BY RONAK AND MEHRDAD

As biomechanists and ergonomic designers of medical devices, we understand the importance of bringing our experiences and whole selves into new scenarios, especially when designing products for individuals with diverse needs and wants. Our experiences in healthcare practice and management have led to wonderful discoveries in burgeoning practice, research, and teaching in human factors and work design.

We believe that there is still much to discover about the rich contributions of humans in a working system, and when designed well, this orientation can optimise health. This is particularly vital in healthcare, where health is central to performance for both consumers and the workers – the administrative and clinical service providers, the supporting staff, the leadership, and the governance teams.

In designing medical devices, it is crucial to consider the diverse needs and wants of the individuals who will use them. By incorporating human factors and work design principles, we can create products that optimise health and performance for both consumers and healthcare providers. Through our experiences in healthcare practice and management, we have discovered that designing products with a focus on human factors and work design can lead to improved outcomes for all individuals within the healthcare system.

As a result of our experiences, we recognise the importance of designing medical devices that are functional and considerate of human factors and work design principles. This approach can lead to increased user satisfaction, reduced errors, and improved overall performance of the healthcare system.

By understanding the unique needs of healthcare providers and consumers, we can create medical devices that are easy to use, efficient, and effective. This, in turn, can lead to improved health outcomes for consumers, reduced workload and stress for healthcare providers, and increased efficiency and productivity for the healthcare system.

Our experiences in healthcare practice, design, and management have led us to recognise the importance of human factors and work design principles in the development of medical devices. By continuing to prioritise these considerations, we can contribute to improving the healthcare system and, ultimately, the health and well-being of individuals. We are committed to utilising our expertise to create innovative and effective medical devices that consider the unique needs and experiences of healthcare providers and consumers.

ACKNOWLEDGEMENTS

Thanks to Mehrdad Hassani, my husband and co-contributor to this chapter with whom I have discussed flexible practices in the ergonomic design of medical equipment and continuous quality improvement strategies. I am grateful, also, to my parents who were my greatest source of encouragement in pursuing this career path.

REFERENCES

Ahmed, H., & Ahmed, M. (2023). Human systems integration: A review of concepts, applications, challenges, and benefits. *Journal of Economics and Sustainable Development, 14*(4), 30–39.

Cheng, M. (2003). *Medical device regulations: Global overview and guiding principles*. World Health Organisation.

Ducrocq, S. C., Meyer, P. G., Orliaguet, G. A., Blanot, S., Laurent-Vannier, A., Renier, D., & Carli, P. A. (2006). Epidemiology and early predictive factors of mortality and outcome in children with traumatic severe brain injury: Experience of a French pediatric trauma center. *Pediatric Critical Care Medicine, 7*(5), 461–467.

Golmohammadpour, H., Garosi, E., Taheri, M., Dehghan, N., Ezbarami, S. M., & Karanikas, N. (2023). Redesign and ergonomic assessment of a chest-support baby walker. *Work*, 1–11, preprint.

Hall, K. K., Shoemaker-Hunt, S., Hoffman, L., Richard, S., Gall, E., Schoyer, E., Costar, D., Gale, B., Schiff, G., & Miller, K. (2020). *Making healthcare safer III: A critical analysis of existing and emerging patient safety practices*. Agency for Healthcare Research and Quality

Hassani, M., Hesampour, R., Bartnicka, J., Monjezi, N., & Ezbarami, S. M. (2021a). Evaluation of working conditions, work postures, musculoskeletal disorders and low back pain among sugar production workers. *Work*, 1–17, preprint.

Hassani, M., Kabiesz, P., Hesampour, R., Ezbarami, S. M., & Bartnicka, J. (2021b). Prevalence of musculoskeletal disorders, working conditions, and related risk factors in the meat processing industry: Comparative analysis of Iran-Poland. *Work*, 1–17, preprint.

Jin, H., Abu-Raya, Y. S., & Haick, H. (2017). Advanced materials for health monitoring with skin-based wearable devices. *Advanced Healthcare Materials, 6*(11), 1700024.

Lueder, R., & Rice, V. J. B. (2007). *Ergonomics for children: Designing products and places for toddler to teens*. CRC Press.

McDonald, E. M., Mack, K., Shields, W. C., Lee, R. P., & Gielen, A. C. (2018). Primary care opportunities to prevent unintentional home injuries: A focus on children and older adults. *American Journal of Lifestyle Medicine, 12*(2), 96–106.

Minopoulos, G. M., Memos, V. A., Stergiou, K. D., Stergiou, C. L., & Psannis, K. E. (2023). A medical image visualization technique assisted with AI-based haptic feedback for robotic surgery and healthcare. *Applied Sciences, 13*(6), 3592.

Mohammed-Nasir, R., Oshikoya, K. A., & Oreagba, I. A. (2023). Digital innovation in healthcare entrepreneurship. In *Medical entrepreneurship: Trends and prospects in the digital age* (pp. 341–372). Springer.

Mohd Basar, M., Ali, M. F., & Abdul Aziz, A. (2021). Parental differences in knowledge, perception, and safety behaviors regarding home injuries in an urban Malaysian district. *Makara Journal of Health Research, 25*(3), 5.

Webster, K. L., & Haut, E. R. (2024). Human factors and ergonomics in the operating room. In *Handbook of perioperative and procedural patient safety* (pp. 75–86). Elsevier.

Part III

Work Designer Stories

14 Honouring the Healthcare Worker by Championing Good Work Design

Sara Pazell

14.1 AN EMERGING PHILOSOPHICAL ORIENTATION

During my studies for my Bachelor of Applied Science degree in occupational therapy, I made a friend. This friend, Jo Boylan, would, 30 years later, become the co-editor of this Healthcare Insights book. It is fitting that we should get the chance to share some stories because our earliest association was anchored in creating meaningful stories in the daily lives of residents of an aged care facility. Josephine and I shared a role as co-activities directors at an aged care facility in South Australia before I completed my undergraduate degree. I worked a little in Australia, moved back to the United States, studied more, and worked in both countries at various times. I call both America and Australia 'home', though most of my time is spent in Australia now.

As activities directors, we had no interest in being bland or conventional. I was a young occupational therapy university student full of aspiration to make a difference in a world that I believed at the time to be all about community, care, and hope.[1] Altruism was at the core of my being. Still, I was sometimes nervous about the need to follow established procedures.[2] Jo shook that out of me by asking me about how I would like to live if life were measured by mere days or moments. Jo had returned to Australia from her years of living in New York. She was a masterful fitness instructor soon to embark on her early nursing studies. Jo was full of spunk and her energy was unparalleled.

We talked with the residents and asked them to design their ideal activity schedule. Bingo was rarely on the list (though, sometimes, we hosted that in over-the-top fun ways). The activities most of interest to the residents were 10-pin bowling, attending the Aussie-rules football matches at the local stadium, games nights, dress-up

[1] The sentiment of spunk, incessant curiosity, and tenacity derived from my American culture of upbringing was tempered by the simplicity and generosity of Eastern Asian Laos culture. I met a Lao friend in Australia, and I was welcomed into their family as though it were my own. I learned about a new fabric of living woven tightly by the threads of a collective kinship.

[2] These days, I like nothing more than disrupting the norms and being inventive to realise new possibilities.

DOI: 10.1201/9781032711195-17

functions with fashion shows, group fitness, in-house music concerts, face painting, and Chinese New Year dragon parades. The residents spoke, and we finagled the approvals, transportation, medication, and meal regimes to make it happen.

I remember Tony who loved to have his face painted as a tiger character: he had had a stroke, used a wheelchair for mobility, and was hemiplegic. True to character, he would roar and move his hands-mock-paws in the air (often one side with the aid of the other). I recall Max, post-tracheostomy after his throat cancer, who loved to attend the local football. He used an electrolarynx speaking device to communicate, and he did so with gusto. He had his full mental faculties and his own funds, so it was our obligation to fulfil his wishes by purchasing him a few beers at the 'footy'. We were never popular with the nursing staff when we explained his nutritional and hydration intake upon his return from the outing. They had to adjust medication regimes accordingly, but Max was satisfied. Nancy loved the dress-up evenings and was found parading among the rooms of the men when she had been enlivened through an evening of make-up and special attire. Also, Bill, wonderful, happy Bill, whose palliative care was punctuated by his proud display of his many bowling trophies surrounding his bed. He earned these 'wins' when able to play from his wheelchair, using side guards at the lane gutters. He had been enthralled with his ability to participate and compete.

14.1.1 Early Career Activity

I started my early professional career as an occupational therapist working with an adolescent to adult population across several domains, such as mental health outpatient day centre, inpatient acute care, and transitional care. I worked in general hospital-based acute care rehabilitation and step-down transitional units. Also, I provided services in skilled nursing and aged care, and general community care. I practised in Australia and the United States. I was a 'travelling therapist' in the days when that was in demand in the United States. During those times, I developed carve-out clinical niche service programmes. These included wheelchair assessment programmes in aged care settings or neurological and ocular vision rehabilitation services in many settings. In my early 20s, I became a rehabilitation manager in community care and, by my mid-to-late 20s, I commenced master's level studies in business administration and international business. At this time, owing to upheaval in the management of the community care agency that employed me, I was thrust into an acting position of senior management as the executive director of a community healthcare agency. This was a temporary position initially, and the franchisees needed my help. I later learned that they had planned to close the agency, though this is not how things unfolded and my position was soon made permanent. I was a young healthcare executive and while I worked and studied hard, much was navigated by instinct and forged through quality relationships with staff, referring medical practitioners, or representatives of other healthcare organisations.

I revelled in the challenges presented by clinical care delivery, leadership, management, regulatory compliance, and the need for business growth. During these pursuits, I overworked. In my mind, this 'charitable giving' of my time ensured better client and worker experiences. I was intent on learning while ensuring agency compliance and business success. In so doing, I led the California-based organisation through an Office of the Inspector General Federal Bureau of Investigations wedge audit (like the Royal Commission audits that occur in Australia), two Medicare audits, a Joint Commission on Accreditation of Healthcare Organisation audit, and two corporate internal performance audits.

During this time, I also upset the status quo and close-knit political establishment in the county in which we operated. I learned of a plan by the county administration to quietly divest and privatise community home care agency services. This would have created an unfair imbalance of service delivery in the area. I arranged a meeting with a local County Board Supervisor. She was the second one that I had approached: the first had refused to receive me. This situation was news to her. She advised me to question the administration and the board in their next public meeting. I did just that. I spoke during an open public question meeting and asked about these plans. It was a showstopper, and I had no idea of the extent of covert administrative plans that I had uncovered. They agreed to table this issue at the next public meeting, which was standing room only, with full media representation this time. The covert plans were no longer secret. The only way forward was for them to navigate their agreement among board members, and then open this process to public bidding. The course of history was changed, and transparency was required during the transition. We were not successful in the public bid for care (I did not expect that our agency would be successful after causing such unrest), but the dissemination of services was better managed. The monopoly that would have been created in the first instance was disbanded.[3]

During my tenure in home care management, I led a team focused on establishing positive relationships with referrers who wanted a rehabilitation and enablement model of care in community settings: our specialty team included rehabilitation nursing, allied health, and rehabilitation assistant and aide services, forming almost half of our clinical service team. I also studied for a master's in business administration in the evenings during this time which helped shape my management literacy and capabilities. My youth, combined with the opportunity to tackle difficult challenges and learn while being of service to others, fuelled my vigour.

To build the clinical service scope and volume, I reached out to doctors and providers. Importantly, there was a large tender issued to provide 24/7 care support to a young man who was managing his disabilities. He was quadriplegic following a cervical spinal injury that arose from his recent surfing accident. His profile was known; he was a community poster child for safe surfing with a mate in case help was urgently needed. The nursing director and I contacted the consumer to understand his needs, believing that this would provide a more valuable and meaningful way to determine if we could support him. We understood this to be more important than

[3] That same responsive supervisor, when previously a city mayor, pioneered the legislation restricting smoking in public spaces, and the ban of cigarette vending machines in the county. This served as a precedent for the nation and other Western world restrictions. As supervisor, she also led a well-coordinated community response to a massive environmental oil-spill ocean clean-up. She was ready to battle the right issues and she remains a mentor to me today.

trying to present our marketing profile and business capability to the insurer in a competitive bidding process with boardroom presentations.

We contacted the consumer to inform him that we would only contend for the service delivery if we understood, firsthand, what type of support he required and what was most meaningful to him. He invited us to his home. I recall the evening: it was dark, riddled by torrential rains and rife with muddied, pot-hole cavities in the dirt roads for part of the passage. That evening, through nursing support and with family surrounding him, we learned about the person central to the disability and we uncovered aspects of his care needs. He chose to work with us by demanding that the insurer cease all corporate competitive care bidding among external providers, and we were, thus, able to outbid larger and more established organisations to provide his care. The contract was significant: it supported the business and the organisation prevailed. We continued to grow the service delivery in other realms, like private pay nursing aid services and rehabilitation-based home-enablement programmes, which were strategic points of difference from other conventional providers of compensatory care models.

In another healthcare management role, I worked as an administrator at a niche corneal ophthalmology and minor orthopaedic procedure outpatient surgical centre. During this time, I helped the centre achieve accreditation per the Accreditation Association for Ambulatory Health Care (AAHC). Inspired by these incentives in leading practice, we launched innovative projects. They were driven by the ideas from the staff, their daily experiences, interactions with the patients, and patient feedback. We established new methods to reduce waiting room time and congestion. Regular community education programmes were provided in our evenings. Outreach and transportation were provided to support minorities (like agricultural workers, who were a Hispanic population). We introduced surgical care quality measures, like reducing interoperative time to reduce infection risks. Supply chain integration methods were improved with just-in-time supply delivery for most routine supplies. Also, we liaised with anaesthesiology teams to improve team and client communication. I enjoyed the mix of clinical service delivery while meeting business objectives with quality and efficiency metrics. I liked setting up systems that considered human variability in cognition, planning, and performance.

During the healthcare administration and management roles, I was selected to sit on a leadership committee sponsored by the local council. This provided a pathway to learn about innovation. Each month, business leaders on this committee were invited to learn about leading organisations within the county. In this, we could appreciate the business acumen and creativity of the people within our community. We were introduced to sometimes-invisible champions of change and diverse thinking. We met entrepreneurs who innovated (sometimes through constraint-based motivations, and other times because they held strong convictions and aligned their work with their values). I learned about a printing company that evolved to develop what are now commonplace but were new at the time: graphic-designed wine labels to depict mood, emotion, and brand identity, rather than formulaic descriptions of the wine composition (this revolutionised wine marketing and promotional sales). I learned about an agricultural company that implemented phenomenal ergonomic strategies to build efficiency in operations and mitigate ill-health effects from work demands on a migrant worker population. I walked through an architectural firm whose commitment to environmental sustainability and trust in employees foretold what later would become known through accreditation measures like the 'Green Building'[4] initiatives in their building design. They employed activity-based and agile work strategies.[5]

Learning from this model of cross-industry sharing of ideas, I brought this into business management experiences. To support quality improvement, I invited team members of the surgical centre to work with others from different work areas, cultural influences, levels of authority or seniority, and age. Small group 'study success' project teams were formed. The goals were to observe successes and to learn. The mission was to explore and share ideas. Together, the teams could choose another business setting and focus their attention on success. For example, one group studied a hairdresser who greeted customers by name, served refreshments, offered some aromatherapy of choice with a fragrant essential oil, and provided a head massage mid-service. Customers were impressed by these health-oriented, intimate, mind–body gestures.

The 'study success' teams reported to the larger innovation group, including administration management and other project teams. There was no mandate that an idea had to be adopted, just that clever ideas were shared, and innovative design-led thinking was fostered. This underpinned the quality improvement activities but, equally important, connections were made among team members who might not usually get the chance to work together, thus strengthening the social fabric of work.

These experiences stimulated the emergence of my work and eventual PhD studies in human factors, ergonomics, and work design strategy (though these studies were also part of my undergraduate and master's degrees). Also, I continued to attend relevant continuing education and certification courses throughout my working life. It was during my tenure in home and community care management that a friend and colleague brought to my attention the work of her husband. My friend was an occupational therapist, and she was blind. I respected the accommodations that would have been required, especially during the time of her training, to accomplish university education that catered mostly to those without vision impairment. She explained to me her husband's work in behavioural architecture. It meant developing an understanding of the psyche of people and their responses to the built environment: how people behave differently given the design of a structure, or how design of artefacts could, in turn, influence the psyche or alter tactics. I was hooked. I wondered how and why this type of nexus among human health, behaviour, and design strategy was not better known to me in terms of undergraduate career paths.

---

[4] For example, LEED Rating system. www.usgbc.org/leed [accessed 3 Nov 2022]

[5] They had natural lighting, airflow, and ambient temperatures reflective of the season to support circadian rhythms. They were known for greywater use, bike storage facilities, and carpool rewards programmes. They designed their facilities for activity-based work areas; and provided ergonomic furnishings with employees given the freedom to customise workstations to suit their needs, coupled with ergonomics education to inform their decisions. Also, they enjoyed cowbell ringing festivities at the reception desk to celebrate new projects that sustained their business.

Now, I find that I can leverage my knowledge and interests in these realms through the broad applications of human factors and ergonomics in work design. I am fascinated by the construction of health through human factors and systems of work. This includes its governance and leadership systems, the design of jobs, product interaction, technology, and environments. I spend a lot of time thinking about design-thinking: the metacognition about how people design, their orientation to innovative ideas and openness to innovation, and their creative approaches in design. I am interested in design strategy and the restlessness that comes from knowing that change is possible while embracing new and novel ways to go about these changes. I had not realised at the time, but the framework for these ideas was developing in my earliest experiences of clinical care and healthcare systems management. Altruistic acts beget altruism: I recognised that my desire to 'do good things' and be charitable was supported by public policy to derive social benefits (Heger & Slonim, 2022) and I permitted these ideals to flow into the realm of design.

Simultaneously, I hold qualifications in strength and conditioning, yoga teaching, exercise science, fitness, and Thai massage. I studied Thai massage with an amazing team in Chiang Mai: A Tai-Yai village-trained practitioner who had been a monk for decades, and his anatomy-proficient New York wife. They taught the harmony of energy balance combined with physiological and anatomical sciences. I practised these arts throughout my professional studies and work. Yoga instructs me in the need to regulate the way that we breathe, move, stabilise, rest, and think. This is influenced by philosophy, values, and the activation of the autonomic nervous system during movement, poses, relaxation, and meditative activities (Streeter et al., 2012). These practices were found useful among healthcare workers in managing occupational stressors in the COVID-19 era (Zhang et al., 2021). General strength-based and cardio exercise are an outlet also, albeit a different mode of somatic neural activation. Sports and exercise participation leads to stress reduction, anxiety management, and general improvements in mental health (Sharma et al., 2006). These interests have been embedded in my understanding of neurological regulation, systems, product interface, environmental design, and performance. Considerations can be made for an individual or at a macro level for populations of workers. I see the interrelationships swirling in my mind and they have inspired my belief in the magic of design and the marvels of creation and ideation.

14.2 WORK SYSTEM DESIGN

My interests in biological, psychological, and cognitive health in occupational therapy are well aligned. These beliefs are anchored in an understanding of the person, environment, 'occupations', and performance (Baum et al., 2015). The focus is on enabling people to perform at their best and engage in activities that are most meaningful to them. This forms a pathway towards self-actualisation, health, and well-being. The distinction is important because it is not a reductionist medical model that might focus on individual treatment, disease pathology, or risk containment. I did not know the term at the time, but my early study and career experiences and this philosophical orientation were the precursors to my appreciation of *salutogenesis* (Mittelmark et al., 2017; Antonovsky, 1979). Salutogenesis is a philosophy founded on survival, yet it is appreciative of the aspirations to thrive. The term means the 'genesis of health' (Antonovsky, 1979). It is a study of health states and the means to construct health, and, as such, it is inventive, not just preventive. The ideas are not bound by the limits of treatment intervention. The fundamental tenets of salutogenesis focus on helping people achieve a sense of coherence so that their lives are meaningful, manageable, and comprehensible (Mittelmark et al., 2017; Antonovsky, 1979). These ideas are mirrored in positive psychology, organisational psychology, and job design (Knight & Parker, 2019).

My work extends beyond healthcare now, and I apply these fundamental principles in a variety of other industries too, like mining, infrastructure, energy, transportation, manufacturing, retail, education, and so on. In most of my work, I consider macroergonomics: the design of systems and the performance of people in these systems. Human factors and ergonomics take a macroscopic view to consider cohorts of workers and work systems, and how they might work best (Carayon et al., 2015; Shorrock, 2016). We ask:

- Have humans been considered in the life cycle of the system design? (In healthcare, this can mean workers, maintainers, and care-recipients).
- Do we understand business objectives and values, and the expectations on human performance?
- Is the system resilient, regenerative, and agile to sustain well living?
- Is the system established to anticipate, learn from, and respond well to anticipated and new or unknown circumstances?
- Does the system provide the design affordances that enable humans to be well-informed, make the best possible decisions, and take the most prudent action given variable circumstances and fluctuating needs?
- How do humans respond to their social, technical, and built environment?
- What is the reality of their work versus what is imagined, designed, prescribed, or disclosed?
- Are humans understood to contribute to ongoing innovation and solutions-based problems?
- Are the design philosophies clearly articulated, concepts well-aligned, and strategies sound and realistic?

Importantly, ergonomics is a design practice and process, not a product (like an 'ergonomic chair') and not a risk category (like an 'injury risk', a 'financial risk', or a 'reputational risk', for example). It is also not a catch-all term for a fragmented part of performance (like biomechanics alone). It reflects design research, teaching, and practice that relates to human performance with physical, cognitive, psychosocial, and organisational components (Karanikas et al., 2021).

14.2.1 System Resilience

A work system, like the human body and cognitive functions, is not finite or static, but constantly erratic with ever-changing demands in the sociopolitical, environmental, and technical worlds (Carayon et al., 2015). Variability is certain within a complex system (Carayon, 2006). A conventional risk-based management approach might examine a linear cause-and-effect relationship. A system perspective appreciates the complex interplay of factors that add to or detract from performance. When the systems begin to drift from expected performance, this must be noted. The events are understood to arise from poorly regulated or insensitive work systems (Hollnagel et al., 2013). There is a messy reality of work because of the influence of social dynamics (Carayon et al., 2015). At times, we break the rules, yet work improves or, conversely, we follow the rules and get poor outcomes (Hollnagel et al., 2015) because of new and unimagined scenarios.

A sociotechnical systems perspective (Eurocontrol, 2009) with resilience engineering (Hollnagel et al., 2015) acknowledges the interplay of humans, work and work conditions, the environment, and time. There are many interacting factors. System resilience is well established when it can accommodate and deflect stressors. In this sense, it is not the stressor that is the cause of a problem, but the system's ability to detect the stressor and respond. This includes the ability to determine a new or more appropriate path for detection and response in time for the effect to be useful. This is akin to the correct triage system in an emergency department when timely and quality intervention saves lives.

The arrival of a new patient (a stressor) is not the concern. The ability of the system to avoid ambulance ramping, triage patient care needs, attend to the patient with an effective emergency response, and save a life while not compromising others, is the concern.

A sociotechnical approach will try to unify work as imagined, as disclosed, as done, and as prescribed (Shorrock, 2016). The integration of human systems ensures that human-related activities are considered during system planning, design, development, and evaluation (Burgess-Limerick, 2020). This includes recruitment, training, staffing models, worker cohorts, occupational health, safety, human factors engineering, and environmental planning. Safety-critical tasks are important to consider when focusing on business objectives, equipment, and human work requirements (Earth Moving Equipment Safety Round Table [EMESRT], date unspecified). These perspectives anticipate change and consider humans as central to successful operations. A viable system (Beer, 1972) manages routine operations and uses intelligence and smart governance to evolve. This involves the following:

1. Routine performance with existing constraints and resources.
2. Maximising capabilities to leverage existing resources and achieve more or do things better.
3. Envisioning the potential realities if there were added resources.

Beer (1972) also considered the need for intelligence and learning, like through modelling, simulation, research, and good governance (Table 14.1). Beer is known by this quote: '*If it works, it is out of date*'.

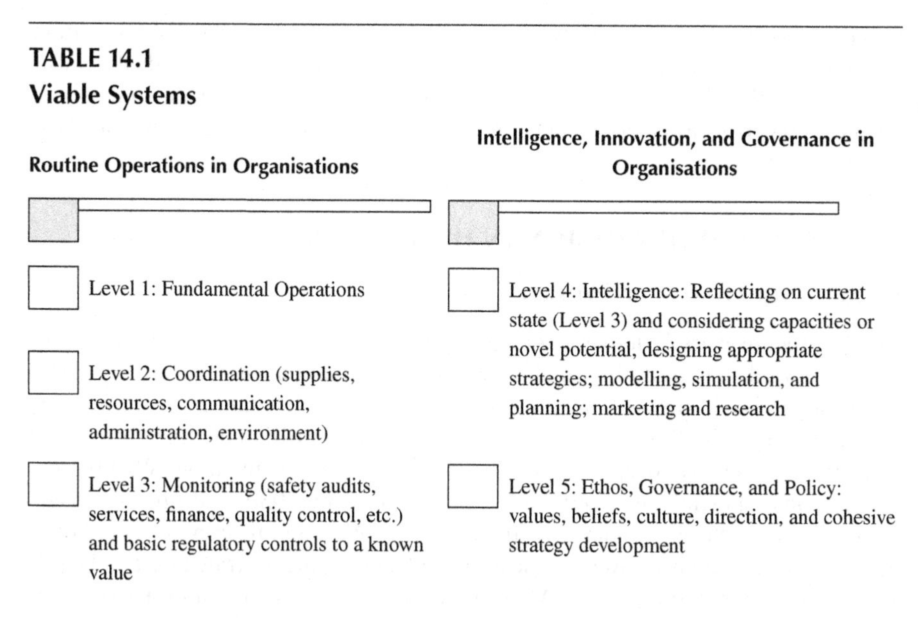

TABLE 14.1
Viable Systems

Routine Operations in Organisations	Intelligence, Innovation, and Governance in Organisations
Level 1: Fundamental Operations	Level 4: Intelligence: Reflecting on current state (Level 3) and considering capacities or novel potential, designing appropriate strategies; modelling, simulation, and planning; marketing and research
Level 2: Coordination (supplies, resources, communication, administration, environment)	
Level 3: Monitoring (safety audits, services, finance, quality control, etc.) and basic regulatory controls to a known value	Level 5: Ethos, Governance, and Policy: values, beliefs, culture, direction, and cohesive strategy development

Source: A simplified adaptation from Beer, S. (1972). Viable systems model. In *Brain of the Firm*. The Penguin Press; Beer, S. (1984). The viable system model: its provenance, development, methodology, and pathology. *The Journal of the Operational Research Society*, *35*(1), 7–25. https://doi.org/10.1057/jors.1984.2

14.2.2 Workplace Service Units

In the workplace, there are business units, such as workforce strategy (human resources), finance, quality management, and procurement. There are operations, health and safety, employee engagement, environment, technology, facilities, and change management services. Business units include engineering, environment, and maintenance services. Rehabilitation and well-being services are factored into organisational design. Marketing and brand strategy, product management, engineering, graphic design, and the like also form part of the social fabric governing work. In a healthcare setting, this may extend to include specialised services or work areas like imaging, pharmaceutical dispensary, medical supplies and equipment, diagnostic centres, clinical records, patient triage areas, treatment rooms, recovery rooms, waiting rooms, rehabilitation centres, storage areas, kitchen facilities, dining areas, cleaning and laundry facilities, staff rooms, and the like. Each of these may have performance needs that theoretically complement one another to advance strategies and values of an organisation. At times, business unit objectives might compete with others. For example, investment in capital expenditure or innovation design trials are costly and these may seem like a short-term budget impediment. Sometimes, safety measures may be viewed as a constraint on operations, and so on.

The balance must be achieved to find the 'just-right degree' of challenge for workers (Lerche et al., 2022). This includes their engagement, motivation, agency, cognitive load, physical demands, understanding of work requirements, autonomy, support, social connectivity, etcetera. They must trust the work system or else risk overriding it to create new adaptations and divest performance (Béguin, 2003). Ideally, business units will align their activities for mutual and harmonised benefit so that core business objectives are met, humans in the system understand the expectations, and values are upheld.

14.3 GOOD WORK DESIGN IN HEALTHCARE

14.3.1 The Healthcare Worker

The healthcare worker delivers essential care and service to those who are sick, ailing, injured, rehabilitating, under medical examination, in need of care coordination, stressed, anxious, needing to adjust to lifestyle changes, or in need of dignity in death. In 2016, there were approximately 59 million healthcare workers worldwide (Joseph & Joseph, 2016). The World Health Organization (WHO) projects a global shortfall of 15 million health workers by 2030, and this is considered a result of chronic under-investment in education and training of healthcare workers, a mismatch between education and employment strategies per health systems and population needs. Also, there are inequities in health access with low and lower-middle income regions, or rural, remote, and underserved areas most affected. Investing in the healthcare workforce advances opportunities for women and youth because of the considerable proportion of women in this industry (WHO, 2022).

The environment of healthcare work is hazardous because of infection risk, chemical exposures, and radiation exposures. There are cognitive distractions, physical demands, shift work, chronic short-staffing, and related stressors (Joseph & Joseph, 2016). The COVID-19 pandemic created a global crisis that led to an unprecedented impact on services. During these times, the psychological burden and wellness of workers came into focus. The pandemic saw 10% of healthcare practitioners, already known to be a cohort at risk of psychosocial distress, experience thoughts of suicide or self-harm in a two-week study period (Bismark et al., 2022). These practitioners experienced significantly higher rates of exhaustion, depersonalisation, depression, anxiety, and post-traumatic stress than their peers. Research continues to show high rates of burnout, stress, anxiety, depression, addictive behaviours, and suicide among workers (Gupta et al., 2021).

In healthcare, we tend to think about the all-important health of consumers as clients, many times at the sacrifice of worker well-being. Yet the health of the workforce is essential to ensure that sensemaking heuristics (spontaneous problem-solving and decision-making) are effective. The staffing resources must be tenable, and the quality of care must be extraordinarily good: the health of internal and external consumers of a system is symbiotic. Burnout and poor well-being are associated with poor patient safety outcomes, such as medical errors (Gibson et al., 2017). Unfortunately, the courts can prosecute an individual owing to their error in healthcare delivery, such as a nurse administrating the wrong medication leading to a patient death. In the case of RaDonda Vaught (NPR, 2022), the courts failed to consider the system of work or dereliction of technology.

Worker engagement, which can be fostered via effective co-design processes, impacts on healthcare errors and adverse events (Janes et al., 2021). Excessive workload, shift work, and long hours are associated with fatigue and reduced quality of patient care (Bolster & Rourke, 2015; Garrubba & Joseph, 2019). Workers must be engaged to the 'just right degree' (Lerche et al., 2022); they must receive meaningful information that can be interpreted readily to make quality decisions and act in a timely and effective manner in known and unknown circumstances. In this way, they can contribute to system performance and make ongoing improvements (Chartered Institute of Ergonomics and Human Factors [CIEHF], 2022). If attentional demands are too low because of automated systems, vigilance decreases (McWilliams & Ward, 2021). If task demands are diminished, engagement reduces, and response resources shrink (like skills, experience, habits, and aptitude); yet a sudden increase in demand can exhaust the system (Young & Stanton, 2002).

While the design of health systems focuses on clinical care, we must remember that the healthcare setting serves a dual purpose. It is also a workplace or 'office' (and, for some, a 'roaming office') where people must do their jobs. The healthcare organisation, in whatever form and shape it takes, is more than a service or retail exchange centre: it represents a workplace for clinical and non-clinical staff. Good design should consider all the features of an effective workspace in which quality, efficient, and accessible clinical care can also be delivered. For example, walking in nature supports active living and can stimulate creativity (Oppezzo et al., 2014). If a walkway in nature is good for healthcare recipients, it is good for workers too.

The healthcare organisation, in whatever form and shape it takes, is more than a service or retail exchange centre: it represents a workplace for clinical and non-clinical staff.

14.3.2 Healthcare Systems and Outcomes

Good work design (Karanikas et al., 2021) is paramount in healthcare and extraordinary care is what we expect: we want our problems solved. In this vein, good work design is either invisible or exceptional but, when it is poor, the consequences can be devastating. We have learned this from the consumer stories in this book. In Australia, there are numerous examples of poor decision-making and system support leading to incorrect or delayed diagnostics causing a critical impact on the consumer. For example, the case of Lucy Dolbel with cervical cancer misdiagnosed for a year before contracting COVID-19 (Barnsley, 2020). There is the case of Naomi Fitzakerley, a mother of three children, whose symptoms of chronic obstructive pulmonary disease was, for years, explained as a fabrication, being 'all in her head' (Swain, 2022). Also, the case of Dean Booth whose skin cancer was cast aside as 'nothing to worry about' until it was too late: an aggressive melanoma was confirmed four months later, and the prognosis was terminal (A Current Affair, 2019).

Research has found that women are more likely to die in intensive care units than men (Modra et al., 2022). This is thought to be owing, in part, because of cognitive biases and practitioner assumptions of clinical presentations to ICU that are more typical of men or women. Further, the dominant sex type of a clinical team can affect their familiarity with and likelihood to detect an illness in the same-sex consumer. As such, system level design changes are considered necessary to combat human error in biased assumptions that can impact on care.

The design of a work system will affect service outcomes. Design must focus on workers' needs and their limitations if they are to do their job well. We must recognise the dynamic nature of activities. Changes are imminent in healthcare delivery: triage and decision-making are constant, and 'flux' is a determinant of the system. The work system must provide channels for learning, and intelligence must exist to interpret the findings. This will permit effective design of new and improved methods of work or working conditions. This requires some freedoms to change and establish 'new rules' for operating.

Changes are imminent in healthcare delivery: triage and decision-making are constant, and 'flux' is a determinant of the system.

The conventions of design help us empathise with workers to understand their health and performance needs and to improve their experiences. Healthcare workers need the right data to inform their decisions, and effective designs of jobs, technology, environments, and work systems. These can regulate the stressors with which they are faced, help them make good decisions, and lead to quality care delivery. This design can and should overlap with positive psychological motivations to support patient or consumer care. For example, quality environmental design can provide for effective wayfinding strategies to reduce patient confusion and anxiety among residents of dementia wards (Golembiewski & Zeisel, 2022), thus, reducing cognitive distractions, interruptions, or general demands on workers.

A positive ageing framework means that aged care residents can be re-enabled to walk and move as much as possible. In turn, this can reduce hazardous manual task exposures for workers associated with patient transfers. Even in acute-care hospital stays, there is a desire to empower patients to be more active, and to design wards to encourage spontaneous movement. Yet, a lack of time and competing work demands are perceived among care staff as a barrier to support patients in their ideal activity levels (Geelen et al., 2022).

The evolution of technology in the design of healthcare systems means that more remote care can be done, and access is less problematic than it once was. Consider that wearable technology designed as jewellery can mediate significant health problems, like diabetes management (Heiss, 2021). Work roles are not diminished by technological advances, but they change what healthcare workers must do and how they must respond to provide quality care (CIEHF, 2022). In healthcare (CIEHF,

2022), there can be automation of administrative and logistical processes, pharmaceutical administration, medical diagnostics, and quality assurance or safety checks. These systems can aid efficiency, reduce errors, improve reliability, and prop valid diagnostic measures. However, the automation can imbalance a system and erode performance, leading to complexity, poor compliance, a lack of integration among subsystems, deskilling, complacency, distrust, and new training or management demands, all of which require competent governance.

14.4 A VISION FOR A CHIEF WORK DESIGN STRATEGIST IN HEALTHCARE

I have a vision of a new carve-out role in organisations within the executive leadership teams: the Chief Work Design Strategist. Imagine this role in the savvy and contemporary healthcare organisation. This organisation recognises the constructive social movement to design work for health. Commonly, they use visual storytelling to evoke emotive and professional experiences (Cortes et al., 2020; Mirkovski et al., 2019). They capitalise on the opportunity to develop regenerative design capability among business units. In this way, routine operations can be complemented by better intelligence and good governance (Beer, 1972). Design capacity, literacy, and capability can emerge in daily practices. The job description to inform the recruitment of such a candidate has been in development in my mind for years. It is time to put pen to paper to express these thoughts. Consider who you might know that could apply for such a role as described below:

14.4.1 The Role of the Chief Work Design Strategist (CWDS)

The advertisement of a Chief Work Design Strategist

The CWDS will bring a scientific, results-driven approach to innovate and solve our real-world problems. They must have a deep understanding of design strategy to influence human performance and business outcomes. They will coordinate design strategies throughout the organisation. The focus must include salutogenic practices to balance pathogenic viewpoints: we want to learn from success as much as we wish to address safety-critical activities in the business. We want to create meaningful, comprehensible, and manageable work that betters our performance. The CWDS will steer the business towards resilience and a sustainable future. They will strive to develop regenerative design capability throughout business units and general operations. The CWDS will also champion the capabilities of conventional designers and manage the development and dynamics of quality design project partners as business suppliers. This will require educating business units on how, when, and why these partnerships can realise added benefits to the way that business is done.

The successful candidate will perform the following:

- Develop and articulate design strategy to integrate business unit objectives. Company values must be upheld while meeting shareholder expectations.

- Manage a small team to help dissect and distil scientific evidence and determine the implications to real-world work practices, which will help identify performance gaps and areas for improvement. The CWDS will articulate the opportunities and problems to be solved through cogent design philosophy, concepts, and strategies.
- Help the organisation embrace ambiguity, iterative design, experimentation, and calculated risk-taking to trial novel approaches in work design.
- Manage 'study success teams' to look outside our industry and learn about new and better ways of work. Form teams from among diverse business partners and operational team members to cultivate a climate of innovation and creativity.
- Manage a significant design strategy budget and portfolio of design projects that best support the vision of the organisation.
- Teach and develop regenerative design literacy, capacity, and capabilities among business partners. Develop the performance metrics that will explain meaningful improvements in the business.
- Develop a culture of storytelling and the mechanisms to communicate these effectively in diverse ways throughout the organisation.
- Work closely with business partners to understand the physical, cognitive, psychosocial, organisational, and environmental factors that enhance or detract from performance among the workforces.
- Work closely with business partners to identify safety-critical work activity that can be supported through design improvements. Learn from case law to support design for 'what-if' scenarios.
- Develop a lifecycle-of-design strategy for our workforce to complement capital equipment and infrastructure investments.
- Work closely with people and performance/workforce strategy teams to design for diversity and enable the inclusivity policies to be realised. This will involve designing for sustainable futures and understanding workforce trends with targeted strategies.
- Be responsible for the development of a diverse, dynamic, quality array of conventional design project partners with whom the organisation can form supplier agreements. They will actively leverage these arrangements in business activities and demonstrate the benefits derived from these relationships.

A persona of a Chief Work Design Strategist

A persona can describe the characteristics of a work design strategist. One is offered that is modelled on me: my experiences, influences, education, interests, and philosophical orientation. Not because these are the only pathways towards the capabilities, but it is a starting point since the ideas are new, and the role does not yet exist in executive leadership teams.[6] Personas are used to help designers understand users and their characteristics (Acuña et al., 2012) (Figure 14.1).

[6] I am aware of design strategists for environmental design or project management. They exist for customer experience and journey mapping, but these refer to external customers, not workers as internal customers. Maybe this role of championing work design exists in an organisation with executive authority (and I am hopeful that it might), but possibly not the same way that is proposed in this chapter.

WORK DESIGN STRATEGIST

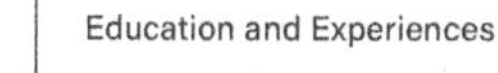

Demographic and Characteristics

- Agitator, comfortable with unpopular and novel ideas
- Can upset the status -quo
- Optimistic
- Risk-taker and innovator
- Hard worker, seeks science
- Seasoned, experienced
- Female, mid-life, worked in male-dominant industries

Education and Experiences

- Human factors and ergonomics, human-centred design
- Macroergonomics and design strategy
- Organisational science and resilience engineering
- Health sciences (physical and phsychocognitive human performance)
- Salutogenic design philosophies
- Business leadership
- Sport and fitness
- PhD-level education and research interests
- Combined clinical, design, and business management experience
- Often not 'fitting in' with conventional role expectations held by others

Pain Points and Frustrations

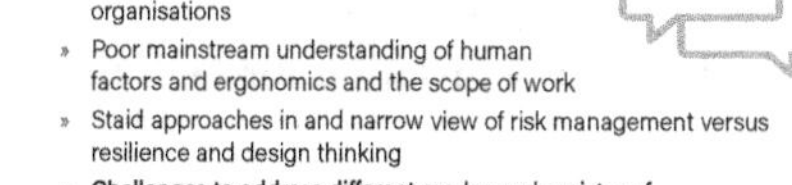

- Need to continually explain the value-proposition of design and design strategy in organisations
- Poor mainstream understanding of human factors and ergonomics and the scope of work
- Staid approaches in and narrow view of risk management versus resilience and design thinking
- Challenges to address different modes and register of communication among varied work cohorts

Needs and Goals

- Creative agency to experiment
- Authority to advance work design strategy
- Trust in process, ambiguity, and iterations in design
- Resources to provide seed-funds for design projects
- Storytelling-compelling, authentic, and emotive
- Responsibility for championing design teams
- Long-term vision
- Manage integrated approaches among business units
- Executive support and influence

FIGURE 14.1 The persona of a work design strategist.[7]

Imagine such a role in healthcare organisations: Healthcare consumer experiences are stories from which we can learn. By involving the clinical and non-clinical workers in the analysis, design ideation, and realisation of the new ways of working and providing care, the workforce become design partners. For example, a collaborative design strategy might focus on the healthcare consumer's exercise capability and functional activities (enablement). If this were in aged care, the expected benefits among residents might be the following:

- Improved activity, social engagement, and life satisfaction
- Improved circulation, mobility, mood, and general well-being
- Fewer hospital admissions
- Fewer hip fractures, bed sores, diabetic complications, cardiovascular distress, or respiratory ailments; reduced all-cause morbidity and mortality

The expected benefits among clinical staff might be the following:

- The ability to partake in altruistic values of care in meaningful ways
- Fewer demands on the need for high-care patient transfers
- Reduced manual task risks during patient transfers
- More accessible work tasks and career options for a diverse array of workers (like older workers and women)
- Improved morale, engagement, and work satisfaction; a better employee experience

[7] Created with a template from a MIRO digital collaborative whiteboard: https://miro.com/app/dashboard/

The expected benefits among non-clinical staff might be the following:

- Positive and dignified work environments
- More accessible environments
- Hope and faith in the caring process associated with ageing
- A better employee experience

The expected benefits to be realised by managers might be the following:

- A more efficient and productive workforce
- Better results per employee engagement and health and well-being measures
- Fewer workplace injuries, insurance claims, or common law expenses
- A broad talent pool from which to derive a more competitive recruitment process and diversity profile of workers
- Discharged obligations to comply with work health legislation through the employee consultation process
- Compliance with healthcare regulatory requirements to prevent or manage illness and support resident well-being through dignified measures
- Improved community goodwill

The work-design strategies could include the following:

- *Leadership:* Embrace philosophies of healthy ageing, salutogenesis, and (re)enablement with the values embedded in board-level literature and activities. Reorient the framework of thinking to diminish ageism biases. Communicate and manage the leading performance indicators that track worker contributions to design discovery, strategy, and realisation.
- *Human Factors and Work Design:* Facilitate workshops to understand the design opportunities for active residents. Conduct task and job role analysis to inform the job-redesign. Develop task-based scenarios for training and storyboarding to communicate strategy. Develop the leading performance indicators of the design strategies and evaluate outcomes. Development of personas, empathy maps, and journey maps to explain design needs and build empathy for work experiences.
- *Operations:* Every job role, clinical or non-clinical, becomes adept at graduating care or physical support to encourage active living and foster independent mobility among residents. Program seasonal sporting or active-participation events that cater to resident abilities. These events require physical preparation and periodisation of training so this can foster an active culture.
- *Quality Care Programmes:* Design daily, weekly, and seasonal activities that are meaningful to residents. Consult and involve residents and their families in the planning of events to suit their interests.
- *Environmental:* Ensure an accessible and attractive space that causes physical activity to occur, such as meandering paths surrounded by nature to warrant exploration by families and residents.

- *Facility Planning:* Create animated digital twin environments with human factors ask-based analysis to plan ongoing design improvements. These may be realised through infrastructure, equipment, supplies, or work activities.
- *Workforce:* To invest in activities-based care providers as part of the healthcare team, including exercise physiologists, occupational therapists, physiotherapists, activities directors, and support staff.
- *Job Design:* To structure jobs that permit rest and recovery, diminish distractions and rushing, and permit the time within the carer role to support active living rather than to 'do things for' residents (that might be seen to be in the interests of time saving).
- *Training:* To develop scenario-based immersive or web-based experiences that provide valid, reliable, and efficient learning mechanisms. These can explain the theoretical underpinnings of the benefits of activity among older residents in simple and memorable ways. They can describe the many ways that this can be encouraged through resident interactions and environmental design. Similar training mechanisms can be provided to non-clinical staff, family members, and visitors, so that they, too, can be part of the solution during their resident interactions.
- *Communication and Messaging:* Interactive messaging that causes movement when learning about the positive and active ageing philosophies. Messaging that is repeated and delivered in novel and unusual ways, embedded in the governance and the promotional materials. Develop compelling human-interest stories with good graphics or digital experiences to evoke an emotional connection to the ideas.
- *Procurement:* Specifications of mobility devices are clearly articulated with purchase agreements well established to ensure a ready supply of quality, effective, and useful technical aids.

Workers should be engaged as part of the solution in the design of good work (Karanikas et al., 2021). This can ensure that design-in-use (Béguin, 2003; Hara et al., 2013) realises the intent of any new strategies.

The work design strategist can establish and oversee the relationships with different conventional designers to include them in agile and well-coordinated ways to work on projects. This means that there are ready suppliers of architecture, interior, industrial, and engineering designers. There may also be need of suppliers of information, user-experience, learning instruction, extended reality, web, information systems, general technology, ergonomics, human factors, and so on.

14.5 SUMMARY

We bring our experiences and whole self into new scenarios and interact with others who have a lifetime of experiences, needs, and wants. I shared a little about my experiences in healthcare practice and management that led to other wonderful discoveries in burgeoning practice, research, and teaching in human factors and work design.

There is much to discover about the rich contributions of humans in a work system and, when designed well, this orientation can optimise health. In healthcare, this is vital, because health is central to performance – for consumers, and for their agents of care (the workers, their administrative and service supporting teams, leadership, and governance). All in this system may wish to meet their potential health capacity given their capabilities, the affordances of their environment, the physical and psychocognitive demands of their work, their culture, their age, and their social supports.

Good health among workers is good for business because there are linkages among occupational health, performance, and clinical outcomes. It makes sense because clear thinking can lead to better decisions and, thus, actions that contribute effectively to the care delivery system. The design of work (in all facets: systems, environments, and operations) can create health opportunities.

This chapter explored the healthcare service experiences underpinned by the persona of a work design strategist. Ideas were shared about good work design in the development of better healthcare systems. A chief work design strategist was proposed as a new executive-level role with authority to influence business in a material way. The implications for this role and this type of thinking can lead to the following:

- The development of regenerative pathways to propagate design activity among all business units. Look to the horizon with long-term visioning of how design can enhance work and care outcomes. Avoid getting 'stuck' in the efforts to perpetually resolve short-term problems.
- Design for what we want, the business or care objectives, and our values; not just for what we do not want. An appreciative, salutogenic perspective can prompt the right considerations about the resources needed to construct resilient systems. Use these strategies to complement the efforts that address safety-critical consequences of work.
- Comfort with ambiguity in which there is continual learning from the realities of work and the applications of design. A threshold tolerance to risk can be set so that clumsiness is permitted if it paves a path to agile and innovative work practices.
- Design of good work for healthcare practitioners leads to good patient outcomes. A focus must be on worker experience as internal customers, as much as on clinical care for external customers.
- Expansive design that considers the system in which care must occur: the environmental design; the nature of the work; the job and requirements of the role; the demands that are physical, psychosocial, and cognitive in nature; the tools, equipment, and technology with which humans interact; and the knowledge and capabilities of the workers.
- Design with humility and empathy: try to understand the needs of the users, the maintainers, the internal and external customers, and the objectives that relate to core values, like sustainability, diversity, or positive ageing. Storytelling as part of routine communication.
- Design strategy that forms part of the routine organisational processes so that the efforts surpass those of one project or one programme. Increasingly, well-coordinated design should become the way that contemporary business is done. It is in the fabric of a smart organisation.

- A work system that enables regenerative capabilities, capacities, and resources among business units. Effective design strategies must be imbued in the way that business is done.
- The reliance on quality research and scientific evidence to prop the work design efforts and enable the people involved in work to contribute in meaningful ways.
- A world in which an organisation is known for its design thinking, creative climate, symbiosis, and innovative culture. Conventional, specialised designers must become part of the extended project teams in a coordinated way. When resources are scant, this can provide for new and agile ways of operating for sustainable service delivery.

ACKNOWLEDGEMENT

Thank you to Elise Crawford, a friend and colleague, with whom I've conversed about resilient practices in work design and the strategies that lead to a sustainable future. She and I are working on the next book writing venture to examine good work design strategies.

REFERENCES

Acuña, S. T., Castro, J. W., & Juristo, N. (2012). A HCI technique for improving requirements elicitation. *Information and Software Technology*, *54*(12), 1357–1375.

Antonovsky, A. (1979). *Health, stress, and coping*. Jossey-Bass Publishers.

Barnsley, W. (2020, July 1). Sunshine coast mum with cervical cancer was misdiagnosed for a year before contracting COVID-19. *Queensland News: 7 News*. Retrieved November 17, 2022, from https://7news.com.au/news/qld/sunshine-coast-mum-with-cervical-cancer-was-misdiagnosed-for-a-year-before-contracting-covid-19-c-1137125

Baum, C. M., Christiansen, C. H., & Bass, J. D. (2015). The person-environment-occupation-performance (PEOP) model. In C. H. Christiansen, C. M. Baum, & J. D. Bass (Eds.), *Occupational therapy: Performance, participation, and well-being* (4th ed., pp. 49–56). SLACK Incorporated.

Beer, S. (1972). *Brain of the firm*. The Penguin Press.

Beer, S. (1984). The viable system model: Its provenance, development, methodology and pathology. *The Journal of the Operational Research Society*, *35*(1), 7–25. https://doi.org/10.1057/jors.1984.2

Béguin, P. (2003). Design as a mutual learning process between users and designers. *Interacting with Computers*, *15*, 709–730.

Bismark, M., Scurrah, K., Pascoe, A., Willis, K., Jain, R., & Smallwood, N. (2022). Thoughts of suicide or self-harm among Australian healthcare workers during the COVID-19 pandemic. *Australian and New Zealand Journal of Psychiatry*, *56*(12), 1555–1565. https://doi.org/10.1177/00048674221075540

Bolster, L., & Rourke, L. (2015). The effect of restricting residents' duty hours on patient safety, resident well-being, and resident education: An updated systematic review. *Journal of Graduate Medical Education*, *7*(3), 349–363. https://doi.org/10.4300/JGME-D-14-00612.1

Burgess-Limerick, R. (2020). Human-systems integration for the safe implementation of automation. *Mining, Metallurgy & Exploration*, *37*, 1799–1806. https://doi.org/10.1007/s42461-020-00248-z

Carayon, P. (2006). Human factors of complex sociotechnical systems. *Applied Ergonomics*, *37*(4), 525–535. https://doi.org/10.1016/j.apergo.2006.04.011

Carayon, P., Hancock, P., Leveson, N., Noy, I., Sznelwar, L., & van Hootegem, G. (2015). Advancing a sociotechnical systems approach to workplace safety – Developing the conceptual framework. *Ergonomics*, *58*(4), 548–564. https://doi.org/10.1080/00140139.2015.1015623

Certified Institute of Ergonomics and Human Factors (CIEHF). (2022, April 25). *Human factors in highly automated systems: A white paper*. CEIHF. Retrieved November 14, 2022, from file:///C:/Users/rego/Downloads/HF-in-Highly-Automated-Systems%20(4).pdf

Cortes, A., Verbrugge, L. N., Sools, A., Brugnach, M., Wolterink, R., van Denderen, R. P., Candel, J. H., & Hulscher, S. J. M. (2020). Storylines for practice: A visual storytelling approach to strengthen the science-practice interface. *Sustainability Science*, *15*(4), 1013–1032. https://doi.org/10.1007/s11625-020-00793-y

A Current Affair. (2019, May 19). Dad told new freckle "all good" before terminal diagnosis. *9 News*. Retrieved November 17, 2022, from www.9news.com.au/national/a-current-affair-medical-doctor-misdiagnose-patients-call-for-reform-latest-news-australia/efa2ad10-1cc0-477b-a98b-ffc9c1596b00

EMESRT (Date unspecified). *Control framework overview*. https://emesrt.org/control-framework

Eurocontrol. (2009). A white paper on resilience engineering for ATM. *Eurocontrol*. www.eurocontrol.int/sites/default/files/2019-07/white-paper-resilience-2009.pdf

Garrubba, M., & Joseph, C. (2019). *The impact of fatigue in the healthcare setting: A scoping review*. Centre for Clinical Effectiveness, Monash Health.

Geelen, S. J. G., Giele, B. M., Engelbert, R. H. H., de Moree, S., Veenhof, C., Nollet, F., van Nes, F., & van der Schaaf, M. (2022). Barriers to and solutions for improving physical activity in adults during hospital stay: A mixed-methods study among healthcare professionals. *Disability and Rehabilitation*, *44*(15), 4004–4013. https://doi.org/10.1080/09638288.2021.1879946

Gibson, D. K., Sampson, A. J., Gembarovski, A., & Newnam, S. (2017). *Linking worker health and safety with patient outcomes*. Retrieved November 1, 2022, from https://research.iscrr.com.au/__data/assets/pdf_file/0006/1321719/Evidence-Review_Linking-worker-health-and-safety-with-patient-outcomes.pdf

Golembiewski, J. A., & Zeisel, J. (2022). Salutogenic approaches to dementia care. In *The handbook of salutogenesis*. Springer. https://doi.org/10.1007/978-3-030-79515-3_48

Gupta, N., Dhamija, S., Patil, J., & Chaudhari, B. (2021). Impact of COVID-19 pandemic on healthcare workers. *Industrial Psychiatry Journal*, *30*(Suppl 1), S282–S284. https://doi.org/10.4103/0972-6748.328830

Hara, S., Shimada, S., & Arai, T. (2013). Design-of-use and design-in-use by customers in differentiating value creation. *CIRP Annals*, *62*(1), 103–106. https://doi.org/10.1016/j.cirp.2013.03.080

Heger, S., & Slonim, R. (2022, January). *Altruism begets altruism* (CESifo Working Papers). Munich Society for the Promotion of Economic Research – CESifo GmbH. ISBN: ISSN 2364-1428.

Heiss, L. (2021, September 17). Designing wearable technologies to solve some of our most pressing health challenges. *Lens*. Retrieved November 18, 2022, from https://lens.monash.edu/@design-architecture/2021/09/17/1383770/designing-wearable-technologies-to-solve-some-of-our-most-pressing-health-challenges

Hollnagel, E., Leonhardt, J., Licu, T., & Shorrock, S. (2013). *From safety-I to safety-II: A white paper*. Eurocontrol. Retrieved July 19, 2022, from https://skybrary.aero/sites/default/files/bookshelf/2437.pdf

Hollnagel, E., Wears, R. L., & Braithwaite, J. (2015). *From safety-I to safety-II: A white paper*. The Resilient Health Care Net: Published simultaneously by the University of Southern Denmark, University of Florida, USA, and Macquarie University, Australia.

Retrieved July 12, 2022, from www.england.nhs.uk/signuptosafety/wp-content/uploads/sites/16/2015/10/safety-1-safety-2-whte-papr.pdf

Janes, G., Mills, T., Budworth, L., Johnson, J., & Lawton, R. (2021). The association between health care staff engagement and patient safety outcomes: A systematic review and meta-analysis. *Journal of Patient Safety*, *17*(3), 207–216. https://doi.org/10.1097/PTS.0000000000000807

Joseph, B., & Joseph, M. (2016). The health of the healthcare workers. *Indian Journal of Occupational and Environmental Medicine*, *20*(2), 71–72. https://doi.org/10.4103/0019-5278.197518

Karanikas, N., Pazell, S., Wright, A., & Crawford, E. (2021, July 25–29). The what, why and how of good work design: The perspective of the human factors and ergonomics society of Australia. In F. Rebelo (Ed.), *Advances in ergonomics in design. AHFE (Applied human factors and ergonomics). Lecture notes in networks and systems* (Vol. 261, pp. 904–911). Springer. https://doi.org/10.1007/978-3-030-79760-7_108

Knight, C., & Parker, S. K. (2019). How work redesign interventions affect performance: An evidence-based model from a systematic review. *Human Relations*, *74* (Advanced online publication). https://doi.org/10.1177/0018726719865604

Lerche, A.-F., Mathiassen, S. E., Rasmussen, C. L., Straker, L., Søgaard, K., & Holtermann, A. (2022). Designing industrial work to be "just right" to promote health – a study protocol for a goldilocks work intervention. *BMC Public Health*, *22*(1), 381. https://doi.org/10.1186/s12889-022-12643-w

McWilliams, T., & Ward, N. (2021). Underload on the road: Measuring vigilance decrements during partially automated driving. *Frontiers in Psychology*, *12*, 631364. https://doi.org/10.3389/fpsyg.2021.631364

Mirkovski, K., Gaskin, J. E., Hull, D. M., & Lowry, P. B. (2019). Visual storytelling for improving the comprehension and utility in disseminating information systems research: Evidence from a quasi-experiment. *Information Systems Journal (Oxford, England)*, *29*(6), 1153–1177. https://doi.org/10.1111/isj.12240

Mittelmark, M. B., Sagy, S., Eriksson, M., Bauer, G. F., Pelikan, J. M., Lindstrom, B., & Espenes, G. A. (Eds.). (2017). *The handbook of salutogenesis*. Springer.

Modra, L. J., Higgins, A. M., Pilcher, D. V., Bailey, M. J., & Bellomo, R. (2022). Sex differences in mortality of ICU patients according to diagnosis-related sex balance. *American Journal of Respiratory and Critical Care Medicine*, *206*(11), 1353–1360. https://doi.org/10.1164/rccm.202203-0539oc

NPR (2022, May 13). Tennessee nurse convicted in lethal drug error sentenced to three years' probation. *Health News from NPR*. Retrieved November 14, 2022, from www.npr.org/sections/health-shots/2022/05/13/1098867553/nurse-sentenced-probation

Oppezzo, M., & Schwartz, D. L. (2014). Give your ideas some legs: The positive effect of walking on creative thinking. *Journal of Experimental Psychology. Learning, Memory, and Cognition*, *40*(4), 1142–1152. https://doi.org/10.1037/a0036577

Sharma, A., Madaan, V., & Petty, F. D. (2006). Exercise for mental health. *Primary Care Companion to the Journal of Clinical Psychiatry*, *8*(2), 106. https://doi.org/10.4088/pcc.v08n0208a

Shorrock, S. (2016, December 12). The varieties of human work. *Humanistic Systems*. Retrieved November 14, 2022, from https://humanisticsystems.com/2016/12/05/the-varieties-of-human-work/

Streeter, C. C., Gerbarg, P. L., Saper, R. B., Ciraulo, D. A., & Brown, R. P. (2012). Effects of yoga on the autonomic nervous system, gamma-aminobutyric-acid, and allostasis in epilepsy, depression, and post-traumatic stress disorder. *Medical Hypotheses*, *78*(5), 571–579. https://doi.org/10.1016/j.mehy.2012.01.021

Swain, S. (2022, November 17). Naomi was told illness was "in her head": She was diagnosed with one of Australia's most common killers. *9 News*. Retrieved November 17, 2022, from www.9news.com.au/national/copd-woman-diagnosed-with-chronic-obstructive-pulmonary-disease-australias-fifth-most-common-killer-exclusive/6c3cbff7-4c4a-4ac9-96da-ada233e7667c?app=applenews

World Health Organisation (WHO). (2022). *Health workforce*. Retrieved November 1, 2022, from www.who.int/health-topics/health-workforce#tab=tab_1

Young, M. S., & Stanton, N. A. (2002). Attention and automation: New perspectives on mental underload and performance. *Theoretical Issues in Ergonomics Science*, *3*(2), 178–194. https://doi.org/10.1080/14639220210123789

Zhang, M., Murphy, B., Cabanilla, A., & Yidi, C. (2021). Physical relaxation for occupational stress in healthcare workers: A systematic review and network meta-analysis of randomized controlled trials. *Journal of Occupational Health*, *63*(1), e12243. https://doi.org/10.1002/1348-9585.12243

15 Sensational Work
The Importance of Sensory-Based Design in Healthcare Service Delivery

Sara Pazell, Pamela Meredith, and Anita Hamilton

We don't see things as they are, we see them as we are.

– Anaïs Nin (Lewis & Nin, 1962)

15.1 A SENSORY WORLD AT WORK

When the sensory needs and preferences of users are considered in the design of work and workplaces, healthcare consumers, practitioners, and educators can be healthful and well. Healthcare workers and consumers can relax more, make better decisions, perform better, and function well (Duarte et al., 2019; Tanja-Dijkstra & Pieterse, 2011). The evaluation of individual sensory profiles is an emerging practice (Dunn, 1997; Yap et al., 2022) and can be used by design teams to consider the varied ways that people might encounter their worlds: through their sight, hearing, touch, kinaesthesia (awareness of the movement of the parts of the body), smell, or taste (Metz et al., 2019). This can be true for any person in any task or environment, but it is especially important in healthcare settings when the users are consumers, such as clients, residents, patients, and family members, or their service providers. It is now well-recognised that consumers may experience trauma (Bradley et al., 2022; Matthew et al., 2022), anxiety (Levy et al., 2007), discomfort, or stress (Duarte et al., 2019) about their healthcare status. They may be tired, in pain, and concerned about their welfare and their future, or those of their loved ones. In addition, the service staff can be exhausted (Al-Omari et al., 2019; López-Cabarcos et al., 2019) or vicariously traumatised by their work exposures (Barnhill et al., 2019), and these experiences can be magnified by skills and staffing shortages (Brzostek & Domagała, 2018).

15.2 THE IMPACT OF OUR SENSES ON PERFORMANCE

Taking the example of just one of our senses, our sense of smell (our olfaction), demonstrates how our senses can markedly affect our quality of life (Winter et al., 2023). Smell is a primal sense, influencing our sense of taste (Postma et al., 2017). Without it, people can report a reduction in quality of life owing to a lack of

DOI: 10.1201/9781032711195-18

enjoyment from food and, thus, present with nutritional concerns. Without smell, there can be safety concerns (e.g., to detect smoke) (Zaghloul et al., 2017) and there may be ongoing worries about being able to cope with an olfactory disorder (Winter et al., 2023). For vulnerable healthcare consumers in need of quality nutrition, taste and smell functions are important (they influence appetite). However, these sense can be adversely affected by medicines (Schiffman, 2018) or treatments (Postma et al., 2017). For others, the noxious odours in some medical settings can be distressing.

The rationale for using sensory theory to inform design approaches is provided by considering the effects of diminished stress and anxiety and improved relaxation and clarity for healthcare workers and consumers; thus, the impact of care can be greater. For example, the use of favourable aromas and fragrances can mitigate the more noxious odours, enhance a sense of safety and comfort, and incentivise a consumer to nourish themselves well. Humanistic critical care environments have been designed using aromatherapy as a sensory-based design strategy to improve quality of life (Fontaine et al., 2001). Aromatherapy has been employed as an intervention for 'effective medical creation' in intensive care units (Satoki et al., 2020). In paediatric care, aromatherapy has been known to lift moods and ameliorate symptoms of nausea and pain (Weaver et al., 2020).

Noise is another example of the relevance of our senses in healthcare. Distractions in emergency departments can impede the cognitive performance of physicians who must make critical care decisions (Dodds et al., 2015). Staff in intensive care units reported increased stress and decreased confidence in their performance when their work was compounded by noise (Schmidt et al., 2020). Such concerns are a particular issue in neonatal intensive care units, where the consequences of frequent alarms on parental distress and the neonate's developing neurological system have been documented (Giudice et al., 2019; Tokunaga et al., 2019; Wallin & Eriksson, 2009). Similar concerns have been reiterated among staff in operating theatres, affecting their concentration and communication. Yet, in these spaces, researchers have found that music had a calming effect without concerning the staff, although it has been found that some types of acoustic stimuli are more distracting than others (Padmakumar et al., 2016). Speech, often considered a distracting noise source, has been tempered effectively via sound masking strategies (such as by using a pleasurable background noise like music) in the open office plan (Lewitz, 2008). Importantly, noise regulation can reduce 'alarm fatigue'[1] in arenas like cardiac care so that the most important information is conveyed, received, understood, and acted upon promptly by clinicians to ensure patient safety (Solet & Barach, 2012). The positive use of sound therapy, conversely, can be used as 'sound healing', a method of psychotherapy (Park, 2022). Sound therapy has been used to treat tinnitus (a buzzing or ringing in the ears, as though the sound were coming from an external source, yet others cannot hear the same noise) (Wang et al., 2020). It has also been used in sensory integration practices to address behavioural disorders (Sinha et al., 2011).

[1] Alarm fatigue is known as sensory overload when healthcare clinicians are exposed to an excessive number of alarms and the result is that alarms are missed, or their significance is not understood. This is akin to 'the boy who cried wolf' in fairy tales.

Older persons commonly encounter deficits in auditory and visual processing. In addition to the experience of pathologies that reduce cochlear and visual sensory processing, generalised ageing can reduce function along these sensory pathways (Gray et al., 2020). Beyond the impact of ageing, sensory acuity may be impaired during times of heightened stress, anxiety, and illness. Focusing on vision, colour design in mental health environments is influential in wayfinding and placemaking, and quality design can address the monotony and under stimulation of the backdrop of most healthcare environments (McLachlan & Leng, 2021). Modern Russian hospitals leverage design to stimulate the visual and tactile systems with colour solutions to highlight navigation objects. They provide patterns to space, order, and structure elements of navigation. Colour design can emphasise or organise relevant text information so that signage is likely to be seen and understood (Kurmangulov et al., 2020). This is important since the complex task of navigating healthcare environments can place unique demands on the visual system (Kay et al., 2023).

Turning to touch in the healing professions, practitioners are called upon in different ways to have appropriate physical contact with their service consumers: a handshake; a body examination; or a gesture of comfort; like the touch of a hand on the shoulder. Yet, reflection on touch is a topic that has been criticised as being absent in medical care training (Lachance et al., 2018) The hand is a sensory organ with motor qualities: it can sense, and it can seek sensations.

Movement is a behavioural factor to modify access to sensory information, like touch, because it involves arousal, attention, motivation, and motor sets. The hand has gross and fine motor aspects (like gross grasp or fine pincer thumb to finger movements), or sensory facets (like instinctive responses to thermal stimuli and strongholds, or refined senses of two-point discrimination and light touch) (Chapman et al., 1996). Tactile, multisensory interaction scenarios are used in design circles to add to expressive, artistic, and therapeutic contexts of environmental design (Azh et al., 2016). Activity-based work design in office settings[2] has been most successful when the physical, psychosocial, and organisational factors (like job design) were considered in design concepts so that users could wholly use space and place to support their productivity and well-being (Marzban et al., 2022).

There are sensory-informed health-related aspects of architectural design that have evolved from the origins of health needs, such as for sanitation and hygiene. Through the ages, there were also spaces for religious rituals that recognised the need for vital sunshine and water (Collins, 2020). Now, there is evidence to support the use of sensory-informed environmental design to mitigate psychoses when these spaces encourage self-expression and meaningful task-based sensory or bodily engagement (Majd et al., 2020). Biophilia, the use of nature and natural elements, can help people feel more at home and reduce their anxiety; without it, discordance may arise (Grinde & Patil, 2009). For frontline health workers, multi-sensory, nature-inspired, 'recharge rooms' yield reductions in stress (Putrino et al., 2020). For patients in a hospital setting, a study identified four of the most important features

[2] Where users can freely move within the environment to choose an open and accessible workspace based on their task-based needs for collaboration or a quiet, sheltered room for solo, concentrated work, for example.

among respondents, with two of these sensory in nature: having space for overnight visitors; privacy features; personal control of lighting; and personal control of temperature (Casscells et al., 2009).

15.3 SENSEMAKING IN HEALTHCARE

The idea of favourable sensory experiences creating a milieu of health and healing was known in early healthcare delivery (Nightingale, 1859). Elements like 'manipulating an environment to be therapeutic' (Nightingale, 1859, p. 100), ensuring clean air, or providing quality light were deemed important to those who were suffering. Research has shown that healthcare consumers (i.e., patients, visitors, and staff) can benefit from sensory-based experiences designed in beneficial ways at the 'touchpoints' of their service interactions (Tanja-Dijkstra & Pieterse, 2011; Yap et al., 2022). Four dimensions of sensory-based intervention have been conceptualised: physical (ambient conditions, space, and its functional use); social (user interactions, or social density); social symbolism (signs, symbols, and artefacts); and natural (sense of nature, being away, or fascinations with design). In the study by Yap (2022), visual, olfactory, auditory, gustatory, and tactile cues were considered. The findings portend that sensory-based design affects everyone in a system – the care recipients, their families and friends, the healthcare providers, and those supporting healthcare service delivery and maintenance. Given that healthcare systems are pressurised by chronic disease management, ageing populations and complex care needs, resource constraints, and demands for increased efficiency and efficacy, we must be smart about using every and any element that can improve the service delivery and experiences among workers, carers, and consumers. Consumers expect personalised therapies and providers (and designers) are discovering the benefits of participatory practices with the active involvement of these people in their care experiences (Noël, 2017). As such, work systems and environmental designers must design solutions that satisfy user needs while simultaneously being constrained by resources.

Workers in health systems are exposed to a wide array of sensory stimuli, often bright fluorescent lighting, or blue lighting; loud or frequently distracting noises; strong odours; the movement of people, beds, wheelchairs, and equipment indoors; and the movement of foot and vehicular traffic in the immediate outdoor environments. The environment may be sterile, busy, and not appreciative of cultural needs (Yap et al., 2022). Equally, healthcare workers can be exposed to trauma – directly or vicariously (Barnhill et al., 2019). Levels of duress or post-traumatic stress disorders (PTSD) can vary among individuals, relating to one's individual sensory profile and trait anxiety (Engel-Yeger & Dunn, 2011; Turjeman-Levi & Kluger, 2022); thus, it is possible that these sensory profiles may predict those who are prone to PTSD.

While the authors have not seen sensory profile assessment used in healthcare employee recruitment strategies or to determine employee performance plans,[3] it could be important to do so to support the best match of jobs and environmental

---

[3] Colleagues have reported that research on this area is being undertaken by the Israeli Army, though the findings have not been published at the time of this chapter submission.

demands to individuals with different sensory preferences and needs. Similarly, understanding one's own sensory profile can support environmental modifications and development of self-management strategies to improve sustainability in a job. For example, the healthcare worker must sense, interpret, and determine appropriate actions during many patient interactions. As such, they need few distractions and clarity of thought to determine non-verbal cues of patients or consumers (like their body language, facial expressions, tone of voice, or skin colouration and signs of integrity), or to make sense of verbal communication. This experience can be complicated by consumers who have difficulties expressing themselves because of angst, clinical conditions, or cultural and linguistic differences. The workers must rely on the integration of their senses: systems that integrate modalities of touch, sight, sound, taste, smell, temperatures, balance, pain, body positions, movement trajectories, or the physiological condition of the body, mood, and mind to perform well (Clements-Croome et al., 2019). Difficulty with the integration of the senses can make it harder to regulate their moods and emotions, or to respond well to patients, avoiding error, misjudgements, confusion, or rushed actions. Failure to consider these circumstances can result in medication errors (Abdel-Latif, 2016; Jang & Lee, 2016), diagnostic errors (Gray et al., 2021), productivity impacts, reduced job satisfaction (Bhandari et al., 2010), physical strains and pains (McCleane, 1999), and absenteeism (Kandemir & Şahin, 2017; Tumlinson et al., 2019).

As a further consideration, healthcare staff with an understanding of the importance of sensory systems and individual differences in sensory needs and preferences can support the appropriateness of their responses to consumers. For example, if they observe a consumer becoming escalated in mood, they might offer some time in the garden or a dimly lit room away from excessive stimuli. Further discussions or interventions regarding sensory needs may also be useful for these consumers, such as those conducted by occupational therapists.

15.4 SENSORY MODULATION AS A TREATMENT

The idea of sensory modulation as a treatment for trauma has been explored in occupational therapy literature (McGreevy & Boland, 2020; Warner et al., 2013). Active coping strategies that manage sensory sensitivities can minimise distress (Meredith et al., 2015a). The research is encouraging for the use of sensory- and attachment-informed interventions to modulate pain (Meredith et al., 2015b, 2021). If the relationship between occupational psychosocial and physical health and the design of the sensory world is considered, then it behoves employers to create better work experiences through environmental and job design. Employers must enact their commitments to provide inclusive, healthful, and safe work. Section 32 of the Australian Work Health and Safety (WHS) Act (2011) described the legislated offence of a failure to comply with a health and safety duty if a person conducting a business or undertaking is deemed to have a duty of care, fails in that duty, and exposes an individual to a risk of death or serious injury or illness (Work Health and Safety Act 2011, 2011). Occupational stress and trauma exposures, accentuated in those with certain sensory profiles (Machingura et al., 2019; Meredith et al., 2015a), can have deleterious effects on individuals (Cobb, 2022; Guidi et al., 2020; ISO, 2021;

Queensland, Work Health and Safety, 2022). It is foreseeable, then, that design that best supports the sensory needs of users in a system (workers and healthcare recipients) forms part of the synergistic organisational strategies (Sorensen et al., 2019) that are essential to quality care.

A sensory approach to work design of healthcare settings can perform the following:

- Support people in need of emotional regulation to process their sensory world, integrate sensory information, and respond in effective ways in a work system. In turn, this can reduce anxiety, diminish stress, and abate depressive episodes.
- Diminish re-experiencing trauma (Diego, 2011; Littleton & Grills-Taquechel, 2011), to help distinguish the traumatic memories from present mindfulness, and reduce vulnerabilities to repeated trauma or victimisation.
- Diminish hypervigilance that leads to increased anxiety or stress (Littleton & Grills-Taquechel, 2011).
- Reduce avoidance behaviours (Robinson, 2001).
- Support effective coping mechanisms (Jones et al., 2003).

15.5 CREATING 'SENSATIONAL WORKPLACES'

It is thought that the integration of our senses and sensory experiences plays an important role in the experience of workers and can be influenced by the environment, leadership strategies, the organisational framework, the design of jobs, and personal health behaviours. These influences affect interactions with others and, importantly, healthcare consumers, if considered in a healthcare setting. In recognition of this, an emerging research stream on 'sensational workplaces' has been developed for honours research students in occupational therapy,[4] with strong industry support from a work design consultancy practice.[5]

A review of the terms related to the realities of a sensory world can be helpful to consider in work design:

- *Sensory Processing:* Sensory processing refers to the ability to register and modulate, or regulate, sensory information such as touch, taste, smell, hearing, sight, balance, and movement. In other words, sensory processing refers to how the nervous system receives, interprets, and manages sensory stimuli.
  - The term can be defined differently in different fields of practice. For example, in psychology, sensory processing refers to the emotional reaction to sensory stimulation. In occupational therapy, it refers to the task-based behavioural (observable) responses to sensory stimuli that may fall within or exceed varying threshold capacities (Turjeman-Levi & Kluger, 2022). It is deemed that there are threshold tolerances (high or low) and behavioural responses (active or passive) (Metz et al., 2019).

[4] Discipline of Occupational Therapy, University of the Sunshine Coast: www.usc.edu.au/study/courses-and-programs/bachelor-degrees-undergraduate-programs/bachelor-of-occupational-therapy-honours

[5] ViVA health at work: www.vivahealthgroup.com.au

- *Sensory Integration:* The ability of the brain to organise and synthesise sensory information from varied stimuli to develop a comprehensive understanding of the occupational environment and use this information to plan and execute appropriate responses to the sensory world (Clements-Croome et al., 2019; Miller et al., 2007). Sensory integration refers to how we organise the sensory input to respond adequately to life demands (Humphry, 2002).
- *Sensory Processing Disorder:* Disorders can arise and are impacted by genetics, neural development, and environmental experiences. People with difficulty processing sensory inputs may develop a disorder (especially when studied per population norms). These disorders include hypersensitivities (e.g., ambient noises may cause stress), hyposensitivities (e.g., in touch, a person may burn themselves if they are not sensitive to heat or trip if they are not sensitive to uneven surfaces), or difficulty integrating information leading to attention deficits, learning difficulties, and behavioural management needs (Crasta et al., 2020).
- *Sensory Profile:* A sensory profile refers to the measure of traits, sensory experiences, and behavioural patterns related to sensory processing (Dunn, 2001; Machingura et al., 2019). Based on the configuration of a threshold tolerance and the trends in behavioural responses, it can be classified into four quadrants: sensory sensitivity; sensory avoiding; low registration; or sensory seeking (Metz et al., 2019). The profiles can be used as tools to inform a work design strategy, modify, or enhance the environment, determine effective communication methods, or prescribe health behaviours, like suitable exercise types. For example,
  - An individual with high sensitivity to noise (hypersensitivity or sensory sensitivity) can benefit from workplace design with sound-absorbing materials, private office spaces, or sound masking.
  - An individual who has a low registration (hyposensitivity) to sounds (e.g., difficulty discerning their name being called against a backdrop of other sounds) might benefit from work settings with fewer ambient noise distractions or the use of non-sound stimuli to engage attention (like waving one's arms to create visual stimuli).
  - An individual may be restless if they actively seek touch when none can be found in their environment. These individuals might benefit from having tactile cues in the environment, like sensory panels of rough surfaces, knobs that turn, crevices in which to trace one's finger, or water displays in areas that permit touch.
  - Providing varied lighting options, areas with natural lighting, areas with dampened hues in the lighting, or adjustable focal task lighting can accommodate different sensitivities to light stimuli (for some, bright lights or certain hues can be irritants).
  - A person who actively seeks movement can benefit from walking meetings, or action-oriented roles in team meetings, like walking around a room to disseminate material or when speaking, or with the use of an active desk design to sit or stand during activities.

15.6 ADVISING ON 'SENSATIONAL WORK EXPERIENCES': REFLECTIONS OF A WORK DESIGN STRATEGIST (SARA)

In a consultancy practice,[6] the Adolescent/Adult Sensory Profile® (AASP)[7] is a tool with suitable psychometric properties that has been used to examine the traits, sensory experiences, and sensory processing behavioural patterns of workers in different industries and workplaces: finance; law; architectural and design; occupational health and safety; mining control centre operations; computer programming and information technology; aviation; elite sport (coaching and playing); childcare; higher education; and among senior technology and innovation teams. The AASP is a survey instrument comprised of 60 questions about profile sense types and self-rated evaluation of behavioural responses to sensory experiences: touch, taste, smell, hearing, sight, activity, and movement. The purpose of this testing is to gain an understanding of individual sensory processing and underpinning behaviours and preferences (Brown & Dunn, 2002). Notably, a person need not have a recognised disability to register with a unique rating on these scales. It is estimated that up to 15% of the population have more intense sensory patterns than tested norms (Miller et al., 2007; Pletschko et al., 2016), which can affect their behaviour and quality of life (Abernethy, 2010). Excerpts of work design, the 'personal passports', have been anonymised and are provided in Figures 15.1 and 15.2. The anecdotal responses to receiving these 'personal passports to sensational work' have been overwhelmingly favourable. These ethnographic, empirical studies that were embedded in work design discovery projects (Karanikas et al., 2021) suggest that formalised research is needed and, because of the volatility and occupational stress concerns in healthcare (Ruotsalainen et al., 2016), it would be useful to apply such studies in a healthcare setting.

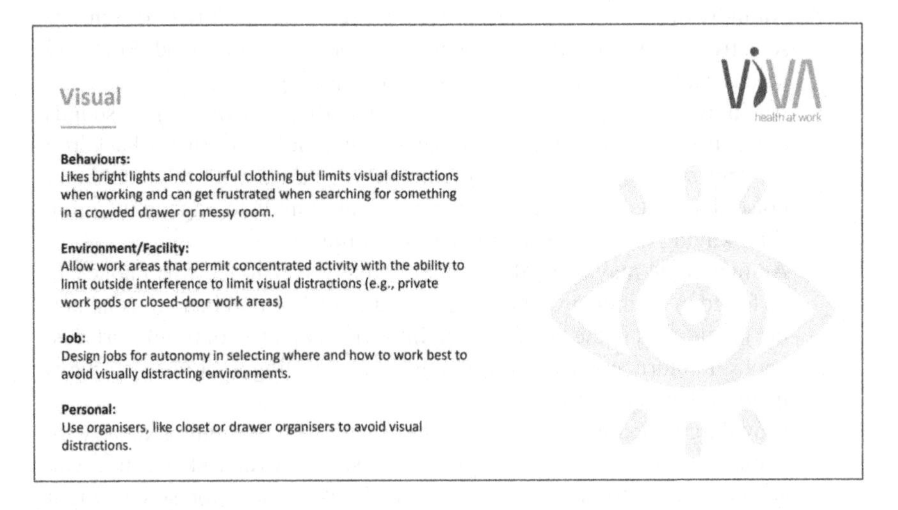

FIGURE 15.1 Sample recommendations for managing visual stimuli in a respondent with elevated sensory-seeking behaviours.

[6] www.vivahealthgroup.com.au

[7] www.pearsonassessments.com/store/usassessments/en/Store/Professional-Assessments/Motor-Sensory/Adolescent-Adult-Sensory-Profile/p/100000434.html

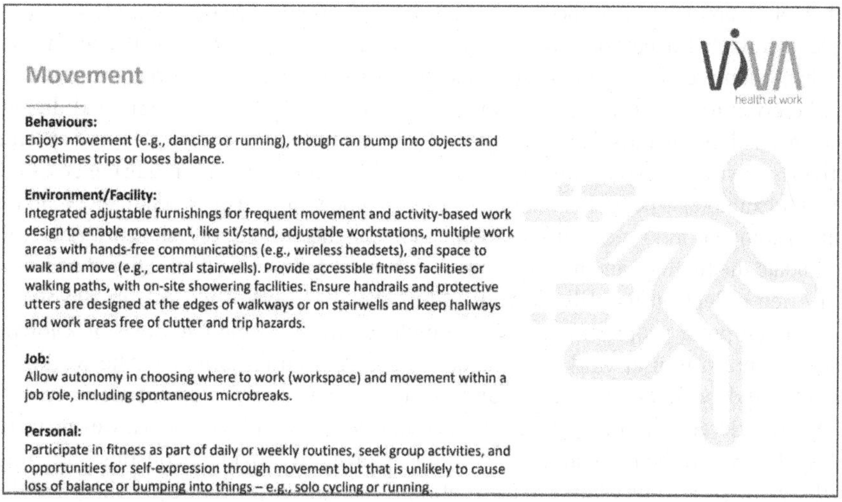

Movement

VIVA health at work

Behaviours:
Enjoys movement (e.g., dancing or running), though can bump into objects and sometimes trips or loses balance.

Environment/Facility:
Integrated adjustable furnishings for frequent movement and activity-based work design to enable movement, like sit/stand, adjustable workstations, multiple work areas with hands-free communications (e.g., wireless headsets), and space to walk and move (e.g., central stairwells). Provide accessible fitness facilities or walking paths, with on-site showering facilities. Ensure handrails and protective utters are designed at the edges of walkways or on stairwells and keep hallways and work areas free of clutter and trip hazards.

Job:
Allow autonomy in choosing where to work (workspace) and movement within a job role, including spontaneous microbreaks.

Personal:
Participate in fitness as part of daily or weekly routines, seek group activities, and opportunities for self-expression through movement but that is unlikely to cause loss of balance or bumping into things – e.g., solo cycling or running.

FIGURE 15.2 Sample recommendations for managing movement needs in a respondent with elevated scores in sensory-seeking behaviours and low registration.[8]

What is most novel to me about conducting sensory assessments in the workplace to inform design is how common it is to discover differing needs among populations of people. That is, most people with whom I have worked or taught in university undergraduate and graduate education and who undertake the assessments have at least one sense type that could sway scores outside the norms. The AASP is one of several tools that can be appropriate for administration to adults in a variety of settings (Dunn, 1997), it is non-intrusive and easy to administer as a survey (especially for English-speaking and literate participants) and can be interpreted per a theoretical model of sensory processing (Dunn, 2001). Our administration of the tool typically occurs via an online survey for our review, evaluation, and response per the 'personal passports to sensational work' (excerpts shown in Figures 15.1 and 15.2). Respondents self-report the frequency of their behavioural response to stimuli (almost never, seldom, occasionally, frequently, almost always) and the administration time is estimated to be under 15 minutes (Brown & Dunn, 2002). Determination of cut scores and classifications are made. There are different cut scores for adolescents, adults, and older adults, but the scores are normally distributed. The classification system describes behavioural responses to sensory processing quadrants as follows[9]:

- Much less than most people: <2% of the study population
- Less than most people: 2% to <16% of the study population
- Similar to (like) most people: 16% to <84% of the study population
- More than most people: 84% to <98% of the study population
- Much more than most people: >98% of the study population

[8] Difficulty with discerning sense types from a backdrop of ambient senses.

[9] A sample of 950 adolescents and adults without known disabilities completed the final version of the AASP to determine the study population.

When our team delves into the nuanced responses that influence these cut-off scores, we typically find that at least one sense type, if evaluated independently of others, might sway results to be more or less than most people. As such, salient advice and recommendations can be given to inform environmental design, job structure, work design, and the pursuits in health behaviours that can help regulate the neural system (like preferred sporting activities, rhythmic solo cycling and running versus group aerobics requiring coordination, for example). The tool is but one in the armament of an assessment battery, and its administration is followed by a one-on-one meeting (face-to-face or remote by video call) to validate the findings and recommendations, or to inform, extend, or refine the ideas.

I often reflect on its utility through storytelling. One of these stories involves a man, a computer programmer, whose employer was at their wit's end of knowing how to support him at work and manage his behavioural outbursts. The workforce strategy and occupational health and safety managers reviewed his performance. They considered dismissing him and terminating his employment. However, fortunately, I received the referral. I was initially referred to provide 'acoustics screening' of the work areas, but when I delved into the early discovery aspects, I learned that only the employee referred for this review was complaining about the sound: no other workers had complaints. This gentleman had been referred to a physician and an audiologist who found no fault with his systems of hearing. This led me to pause and instruct the employer's managers in a novel approach to the assessment: a design discovery approach informed by sensory theory. They approved the administration of the AASP survey and a site visit in which the work site was evaluated concerning its fit for this worker. Also, a meeting was held with the worker to review the survey findings and to interview him about his experience at work. In this case, the worst possible set-up was arranged for the worker who did not discern sounds against a backdrop of noises compared with most people. Notably, while he could not easily discern one unique sound against other ambient noise, his employer asked him to participate in stand-up meetings in an open-plan office.

When we met, I conducted a (privacy index) speech intelligibility screening by reading 100 random words from a corridor hallway, outside a dedicated 'quiet room', and the worker noted how many words he heard. From more than 12 meters away, outside the closed door and down the hallway, the worker could determine more than 30 words, suggesting that the privacy was low and the design or acoustics treatments in this 'private room' were not well-considered (Neumann, 2003). The worker could hear just fine when there were few ambient noise distractions. However, the room that was meant to ensure privacy for a worker in need was not effective at blocking noises because sounds, like spoken words, were easily determined.

When speech understanding is effortful, as can be the case when trying to listen to a focused auditory source against a backdrop of competing sounds, especially recognisable utterances like conversation, higher-order cognitive processing can be impaired (like applying creativity, understanding complex concepts, using systematic decision-making, or brainstorming) (Marsh et al., 2015). This worker's profile, a male with substantial sensitivity to ambient noise and who had difficulty integrating many senses,[10] situated him amongst people likely to suffer significant trait anxiety (Engel-Yeger & Dunn, 2011).

[10] The testing indicated that he had difficulty discerning sound against a backdrop of noise ('low registration' that was much more than most people).

This worker also had significant sensitivity to touch and bright lighting. Nevertheless, he had been assigned headphones by his employer to accommodate his concerns about the noise levels, and the touch of them was distracting. Also, his workspace was positioned in a corner open plan area, directly exposed to the outdoor lighting and reflections from nearby office building sides and rooftops. This was distracting because he was sensitive and actively avoided reflective glare and bright lighting.

The findings of this work design discovery approach meant that a manageable plan could be developed and explained to the supervisors. This skilful computer programmer could be retained since the business was prepared to accommodate his needs. The employer could meet in enclosed work areas or follow up one-on-one in a private room and provide a meeting agenda with minutes to ensure that the meeting content was received well and understood by the worker. In this case, a private solo-use office set-up was necessary and could be considered by the business (whether per work-from-home strategies or accommodated in the office space). Recommendations were made for the improvements to the environmental design and it was recommended that a referral should be made to the appropriate interior designers and specialists in acoustics treatments.

What I dream is that these approaches form part of the more commonplace methods to design for diversity. This case example could occur in any healthcare setting, and exemplify how, in healthcare, such approaches can construct health and mitigate distress in an industry fraught with circumstances that are anxiety-ridden.

15.6.1 Opportunities in Health Practice, Education, and Research

The application of sensory approaches in a workplace context is a new idea. As a result, there is a mountain of research that can be conducted. As health researchers, we seek to employ interventions that meet the basic requirements of being evidence-based (Krueger et al., 2020). In the case of sensory-based design approaches, we are offered an intervention that is supported by evidence of improving mental health and well-being (Scanlan & Novak, 2015; Wright et al., 2020), but relatively little that considers the outcomes when making sensory-based modifications in the workplace. This represents a very clear research need that is beginning to be addressed clinically or in design and warrants a dedicated empirical focus.

15.7 REFLECTIONS OF AN ACADEMIC OCCUPATIONAL THERAPIST (PAMELA)

As human beings, we have an inbuilt default mechanism that presumes that other people experience the world in the same way we do. Although I learned about sensory integration as an occupational therapist, it was after the birth of my second son, who has a sensory processing disorder, that I was dragged into a better understanding of sensory theory that blew this presumption out of the water. I am now reminded daily of just how incredibly differently each of us can respond to our environments, based on our sensory patterns, and just what the implications of that are, in terms of how we manage the stimuli themselves and the impact of the differences on our relationships, our engagement in the world in general, and our engagement in our workplaces.

Since developing awareness of the ways in which we each experience sensory stimuli differently, I have developed what I think of as my 'sensory lens' which assists me to understand what I see behaviourally and emotionally in myself, my family, and the people around me. In our family alone, we have different levels of sensory sensitivity and, while my husband likes the music in the car turned up loud, my son complains that the noise 'hurts' and is easily distressed by the volume. I like to have lots of light around me, while my son likes his living and working spaces to be 'dingey'. None of these preferences are wrong, but we do need to learn to cohabitate. I have an extremely active friend (climbing mountains more than once a week). When I expressed irritation at an ant climbing on me, she commented that she does not even notice ants. She is often under stimulated, based on her low registration of sensory stimuli, frequently seeking movement and touch, while I am often overstimulated, seeking to isolate myself from such stimulating individuals and environments. It is nothing personal, of course, but it can feel like a rejection when an invitation is declined, so applying a sensory lens may just have saved our friendship.

This sensory lens is applicable in every setting in which we live, work, and play. I detest open plan offices in the workplace and I am underproductive in settings where there is a lot of action, movement, and interruption. In contrast, I have an academic colleague who works best with loud thumping music blaring in her office. I do enjoy loud music, but mainly when dancing and not when writing. My office is typically too hot, so I need a fan, whereas one of my other work colleagues has an office that is typically too cold, so she stores various layers of clothing there. I worked with one staff member who insisted that her office temperature be regulated to 24.5 °C, otherwise she could not be productive.

As seen in the above examples, as relatively clever human beings, we are typically good at making adaptations to create environments that are more conducive to optimal performance. It seems self-evident that if one person is cold and another hot in a climate-controlled environment, one should don a jacket and the other use a fan. Yet, in many workplaces, staff are exposed to, and feel forced to tolerate, stimuli with little regard for how they or employers could modify the task or design the environment to better meet their needs. Indeed, I feel confident in suggesting that most employees and employers have not developed a sensory lens for their workplace. A notable exception, now, is in inpatient mental health services for adolescents and adults. Here there is a recognition of the need for control of lighting, quiet spaces, access to referred music, weighted modalities, and a wide range of other sensory equipment and activities to support self-regulation.

With a sensory lens, we can be proactive in making adaptive changes in workplaces, rather than reactively managing situations where people may underperform, and become distressed, disgruntled, and even depressed or anxious, because of levels of sensory stimuli that are contraindicated. We can pose questions: Is this the best job for this person? Are there other roles within this workplace that would be better suited for this person? Are there ways that we could modify the task or the environment that would make this job more suitable/manageable?

When I share my knowledge about sensory processing with others – students, therapists, colleagues, friends – I typically receive the same response: 'Oh, that makes so much sense'. Clinicians always have clients who come to mind and a new understanding of the reasons underlying their behaviour. Applying a sensory lens can diminish and even prevent workplace distress and separation of employees.

For my son, the occupation of student was notably impacted by his sensory processing disorder. Aware of his distress in loud classrooms and playgrounds, inability to process verbal instructions, and discomfort caused by movement around him and teachers with loud commanding voices, it was clear that he was not going to perform optimally in a traditional school setting. Indeed, we observed him to be dissociated (or frozen) in class, and he came home every day completely overwhelmed and frequently 'melting down'. Although we sought alternative educational options, he eventually proactively identified a fabulous school for himself. It was a school with, among other things, small class sizes, choice of seating (chairs, bean bags, therapy balls), no requirement to engage in sport, private spaces to spend time if needed, and encouragement by staff to utilise strategies to remain regulated. Using the sensory strategies, he identified and developed at home and at school, I am now proud to say I am the mother of a university graduate.

While some sensory adjustments can occur automatically, this is not always the case. Whether in a classroom, workplace, or heath setting, if in doubt, it pays to consult an occupational therapist trained in sensory approaches. You know the adage, 'A stitch in time. . .'.

15.8 REFLECTIONS OF AN EDUCATOR, RESEARCHER, AND PRACTITIONER (ANITA)

I am a registered occupational therapist working in higher education. My practice experience and research led me to become interested in the role of Animal Assisted Interventions (AAI), specifically therapy dogs, in supporting the mental health and well-being of students. Research involving dogs and children has been conducted since the 1960s, and therapy dogs are now commonly included in therapeutic and educational settings (Sorin et al., 2015). I have chosen to explore the case study of Cooper the therapy dog in this chapter because this is an animal-assisted sensory-based approach, relevant to many areas of health practice.

When the first COVID-19 pandemic lockdown was announced in early 2020, my husband (a registered psychologist) and I decided that we had a unique opportunity to train a puppy as a potential therapy dog. Successful therapy dogs are calm and gentle; they initiate affection towards humans and respond positively to human affection; they adapt quickly to unfamiliar people, places, sounds, and smells. They are calm in the presence of other (non-aggressive) dogs and never show aggressive behaviour (Sakurama et al., 2023; Skeath et al., 2015). To find our potential future therapy dog, we connected with a trusted breeder and a reputable training organisation, and now, more than three years later, our 'puppy' Cooper is a Level 2 therapy dog working with school and university students. Here's how Cooper uses a sensory-based approach to communicate with the students and staff he 'works' with.

The ability of dogs to detect human attention using eye gaze is important for the role of a therapy dog (Gácsi et al., 2004). One of Cooper's roles in the school is to greet students as they arrive and to support any students having difficulty separating from their parents. Cooper notices when a parent looks towards him, silently seeking support. He returns their gaze and goes to them, creating the opportunity for the parent and the child to engage with him and then transition to the classroom.

Cooper is not simply relying on sight, however. A dog's sense of smell is its ultimate 'superpower'. Dogs use chemo-signalling to understand olfactory information about other animals and their environment. One of the ways therapy dogs use chemo-signalling is to recognise human emotions, such as stress, fear, or happiness (Kokocińska-Kusiak et al., 2021). In his role as a therapy dog in school and university settings, Cooper has a way of finding staff members who are worried or stressed. Using his sense of smell, Cooper locates the 'stressed worker' and sits beside them, gently leaning against their leg while they work. Usually, this leads to the person reaching out and gently patting, or stroking Cooper's head or back.

For most, holding or patting an animal helps them to feel calm, when experiencing stressful emotions (Fine, 2010) by decreasing heart rate and blood pressure (Friedmann et al., 2015). The fur on Cooper's back is soft, wavy, and medium length; you can run your fingers through it. His face is 'open' (no curly hair on his face) so his face is like black velvet. Like many dogs, Cooper invites affectionate touch such as long caresses down his back, scratches on his hips, and gentle massages around his ears. Affectionate touch is calming for dogs and humans and supports emotional co-regulation (Beck, 2015).

As mentioned, Cooper tends to 'lean' on people. By doing this, he is using deep-pressure touch to activate proprioception and to communicate with humans. Recently, a child was sent to the administration building to meet with the principal about her in-class behaviour. The student was sitting on a chair in the reception area, sniffing, and not interacting with any of the staff. Cooper was in another room and detected that the child was distressed. He took himself to the child and simply sat beside her. While still quiet and sniffing, the child reached out and started stroking Cooper's back, Cooper leaned gently on her. Leaning is a form of 'therapeutic touch' (Fine, 2010). One of the strengths of having a therapy dog offer therapeutic touch is that it is a safe way to convey 'acceptance, nurturance, intimacy, safe touch, and physical affection' in a professional setting (Stewart et al., 2013). Cooper was able to help the child regulate her emotions through therapeutic touch.

All seven senses – sight, hearing, smell, taste, touch, proprioception (deep touch), and interoception (recognising one's physiological and psychological state) – influence our health and well-being. Cooper employs all his sensory systems when he works as a therapy dog and through this role, he is sensitive to the sensory needs of students (and staff). Because of this, he helps them regulate their sensory systems, manage their emotions, build relationships, practice social skills, and participate more fully in daily life.

15.8.1 Overview of the Reflections

There are many ways in which having a good understanding of sensory systems and sensory approaches can inform creative ideas for assessments and related interventions or design strategy. Personal stories have been relayed in humble and sensitive ways because developing empathy about sensory needs is a necessary start to these design improvements. Case examples have considered studies on environmental and job design for workers, environmental design for consumers, reflections on educating a child with a sensory processing disorder, and the introduction of dogs to achieve sensory regulation among students. Cooper the dog was mentioned because of his extra superhuman sensory powers of smell and what his presence offers, in turn, to students to help them to regulate their sensory needs. Sensory-informed strategies highlighted in these examples show how we can support people of all ages and in all settings to achieve neurological regulation and function well in their desired and required occupations (and that occupation could be a carer role or it could be the role of a consumer of care, like a healthcare recipient).

15.9 SUMMARY

Human beings are unique and, thus, their environments and service experiences (be it job design for workers or service design for consumers) must accommodate these diverse needs. There are remarkable opportunities to achieve positive experiences when sensory theory is used to inform the design of work. Sensory-based design approaches offer a neurological basis to support high-performance and quality healthcare and health education. This chapter has outlined sensory processing considerations with applications that can inform effective design in health and work settings. With adequate insights, innovative design strategies can be developed to meet the neurological needs of different users in healthcare settings who might otherwise be subject to the disturbances that cause stress and anxiety. Healthcare settings (through the environments, jobs, products, training, educational systems, communication modes, and leadership methods) can be improved by an understanding of the diverse needs and preferences of people exposed to sensory stimuli such as smell, sound, sights, touch, and movement.

REFERENCES

Abdel-Latif, M. M. M. (2016). Knowledge of healthcare professionals about medication errors in hospitals. *Journal of Basic and Clinical Pharmacy*, *7*(3), 87–92. https://doi.org/10.4103/0976-0105.183264

Abernethy, H. (2010). The assessment and treatment of sensory defensiveness in adult mental health: A literature review. *The British Journal of Occupational Therapy*, *73*(5), 210–218. https://doi.org/10.4276/030802210x12734991664183

Al-Omari, A., Mutair, A. A., Shamsan, A., & Mutairi, A. A. (2019). Predicting burnout factors among healthcare providers at private hospitals in Saudi Arabia and United Arab Emirates: A cross-sectional study. *Applied Sciences*, *10*(1), 157. https://doi.org/10.3390/app10010157

Azh, M., Zhao, S., & Subramanian, S. (2016). Investigating expressive tactile interaction design in artistic graphical representations. *ACM Transactions on Computer-Human Interaction*, *23*(5), 1–47. https://doi.org/10.1145/2957756

Barnhill, J., Fisher, J. W., Kimel-Scott, K., & Weil, A. (2019). *Trauma-informed healthcare approaches, a guide for primary care* (pp. 197–213). Springer. https://doi.org/10.1007/978-3-030-04342-1_11

Beck, K. R. (2015). *The impact of canine-assisted therapy and activities on children in an educational setting*. https://fisherpub.sjf.edu/education_ETD_masters/312

Bhandari, P., Bagga, R., & Nandan, D. (2010). Levels of job satisfaction among healthcare providers in CGHS dispensaries. *Journal of Health Management*, *12*(4), 403–422. https://doi.org/10.1177/097206341001200401

Bradley, A. S., Adeleke, I. O., & Estime, S. R. (2022). Healthcare disparities in trauma: Why they exist and what we can do. *Current Opinion in Anaesthesiology*, *35*(2), 150–153. https://doi.org/10.1097/aco.0000000000001094

Brown, C. E., & Dunn, W. (2002). *Adolescent/adult sensory ProfileTM: User's manual*. NCS Pearson.

Brzostek, T., & Domagała, A. (2018). Impact of the quality of management on the general improvement of the healthcare system. *Technology and Health Care*, 1–4, preprint. https://doi.org/10.3233/thc-181513

Casscells, S. W., Granger, E., Williams, T. V., Kurmel, T., May, L., Babeau, L., Boyd, D., Davis, D., Ayine, S., & Thomas, N. (2009). TRICARE management activity healthcare facility evidence-based design survey. *Military Medicine*, *174*(3), 236–240. https://doi.org/10.7205/milmed-d-02-6308

Chapman, C. E., Tremblay, F., & Ageranioti-Bélanger, S. A. (1996). Hand and brain. *Part 5: The Sensorimotor Hand*, 329–347. https://doi.org/10.1016/b978-012759440-8/50022-0

Clements-Croome, D., Turner, B., & Pallaris, K. (2019). Flourishing workplaces: A multisensory approach to design and POE. *Intelligent Buildings International*, *11*(3–4), 131–144. https://doi.org/10.1080/17508975.2019.1569491

Cobb, E. P. (2022). *Managing psychosocial hazards and work-related stress in today's work environment: International insights for U.S. organizations*. Taylor & Francis Group. https://doi.org/10.4324/9781003187349

Collins, J. (2020). *The architecture and landscape of health, a historical perspective on therapeutic places 1790–1940* (pp. 9–18). Taylor & Francis Group. https://doi.org/10.4324/9780429459986-2

Crasta, J. E., Salzinger, E., Lin, M.-H., Gavin, W. J., & Davies, P. L. (2020). Sensory processing and attention profiles among children with sensory processing disorders and autism spectrum disorders. *Frontiers in Integrative Neuroscience*, *14*, 22. https://doi.org/10.3389/fnint.2020.00022

Diego, R. J. S. S. (2011). Healing the invisible wounds of trauma: A qualitative analysis. *Asia Pacific Journal of Counselling and Psychotherapy*, *2*(2), 151–170. https://doi.org/10.1080/21507686.2011.588243

Dodds, P., Xiang, N., Sykes, D., Triner, W., & Sinclair, L. (2015). The effects of noise on physician cognitive performance in a hospital emergency department. *Journal of the Acoustical Society of America*, *137*(4), 2248–2248. https://doi.org/10.1121/1.4920201

Duarte, E., Gambera, D. A., & Riccò, D. (2019, July 3–5). *Health and social care systems of the future: Demographic changes, digital age and human factors* (pp. 16–22). Proceedings of the Healthcare Ergonomics and Patient Safety, HEPS, Lisbon, Portugal. https://doi.org/10.1007/978-3-030-24067-7_2

Dunn, W. (1997). Implementing neuroscience principles to support habilitation and recovery. In C. Christiansen & C. Baum (Eds.), *Occupational therapy: Enabling function and well-being* (2nd ed., pp. 182–232). SLACK Incorporated.

Dunn, W. (2001). The sensations of everyday life: Empirical, theoretical, and pragmatic considerations. *The American Journal of Occupational Therapy*, *55*(6), 608–620. https://doi.org/10.5014/ajot.55.6.608

Engel-Yeger, B., & Dunn, W. (2011). The relationship between sensory processing difficulties and anxiety level of healthy adults. *The British Journal of Occupational Therapy*, *74*(5), 210–216. https://doi.org/10.4276/030802211x13046730116407

Fine, A. H. (2010). Incorporating animal-assisted therapy into psychotherapy: Guidelines and suggestions for therapists. In A. H. Fine (Ed.), *Handbook on animal-assisted therapy* (pp. 169–191). Academic Press. https://doi.org/10.1016/B978-0-12-381453-1.10010-8

Fontaine, D. K., Briggs, L. P., & Pope-Smith, B. (2001). Designing humanistic critical care environments. *Critical Care Nursing Quarterly*, *24*(3), 21–34. https://doi.org/10.1097/00002727-200111000-00003

Friedmann, E., Son, H., & Saleem, M. (2015). The animal – human bond. In *Handbook on animal-assisted therapy* (pp. 73–88). Elsevier. https://doi.org/10.1016/B978-0-12-801292-5.00007-9

Gácsi, M., Miklósi, Á., Varga, O., Topál, J., & Csányi, V. (2004). Are readers of our face readers of our minds? Dogs (Canis familiaris) show situation-dependent recognition of human's attention. *Animal Cognition*, *7*(3), 144–153. https://doi.org/10.1007/s10071-003-0205-8

Giudice, C., Rogers, E. E., Johnson, B. C., Glass, H. C., & Shapiro, K. A. (2019). Neuroanatomical correlates of sensory deficits in children with neonatal arterial ischemic stroke. *Developmental Medicine & Child Neurology*, *61*(6), 667–671. https://doi.org/10.1111/dmcn.14101

Gray, B. M., Vandergrift, J. L., McCoy, R. G., Lipner, R. S., & Landon, B. E. (2021). Association between primary care physician diagnostic knowledge and death, hospitalisation and emergency department visits following an outpatient visit at risk for diagnostic error: A retrospective cohort study using medicare claims. *BMJ Open*, *11*(4), e041817. https://doi.org/10.1136/bmjopen-2020-041817

Gray, D. T., Peña, N. M. D. L., Umapathy, L., Burke, S. N., Engle, J. R., Trouard, T. P., & Barnes, C. A. (2020). Auditory and visual system white matter is differentially impacted by normative aging in macaques. *The Journal of Neuroscience*, *40*(46), 8913–8923. https://doi.org/10.1523/jneurosci.1163-20.2020

Grinde, B., & Patil, G. G. (2009). Biophilia: Does visual contact with nature impact on health and well-being? *International Journal of Environmental Research and Public Health*, *6*(9), 2332–2343. https://doi.org/10.3390/ijerph6092332

Guidi, J., Lucente, M., Sonino, N., & Fava, G. A. (2020). Allostatic load and its impact on health: A systematic review. *Psychotherapy and Psychosomatics*, *90*(1), 11–27. https://doi.org/10.1159/000510696

Humphry, R. (2002). Young children's occupations: Explicating the dynamics of developmental processes. *The American Journal of Occupational Therapy*, *56*(2), 171–179. https://doi.org/10.5014/ajot.56.2.171

ISO. (2021). *ISO 45003:2021 occupational health and safety management – psychological health and safety at work – guidelines for managing psychosocial risks* (pp. 1–25). The British Standards Institution.

Jang, H., & Lee, N.-J. (2016). Healthcare professionals involved in medical errors and support systems for them: A literature review. *Perspectives in Nursing Science*, *13*(1), 1–9. https://doi.org/10.16952/pns.2016.13.1.1

Jones, R. S. P., Quigney, C., & Huws, J. C. (2003). First-hand accounts of sensory perceptual experiences in autism: A qualitative analysis. *Journal of Intellectual & Developmental Disability*, *28*(2), 112–121. https://doi.org/10.1080/1366825031000147058

Kandemir, A., & Şahin, B. (2017). The level and cost burden of absenteeism among health care professionals. *İşletme Bilimi Dergisi*, *5*(2), 1–23. https://doi.org/10.22139/jobs.299082

Karanikas, N., Pazell, S., Wright, A., & Crawford, E. (2021, July 25–29). Advances in ergonomics in design, proceedings of the AHFE 2021 virtual conference on ergonomics in design, USA. *Lecture Notes in Networks and Systems*, 904–911. https://doi.org/10.1007/978-3-030-79760-7_108

Kay, K., Bonnen, K., Denison, R. N., Arcaro, M. J., & Barack, D. L. (2023). Tasks and their role in visual neuroscience. *Neuron*, *111*(11), 1697–1713. https://doi.org/10.1016/j.neuron.2023.03.022

Kokocińska-Kusiak, A., Woszczyło, M., Zybala, M., Maciocha, J., Barłowska, K., & Dzięcioł, M. (2021). Canine olfaction: Physiology, behavior, and possibilities for practical applications. *Animals*, *11*(8), 2463. https://doi.org/10.3390/ani11082463

Krueger, R., Sweetman, M., Martin, M., & Cappaert, T. (2020). OTs' implementation of evidence-based practice (EBP): A cross-sectional survey. *The American Journal of Occupational Therapy*, *74*(Suppl 4). https://doi.org/10.5014/ajot.2020.74s1-po2310

Kurmangulov, A. A., Korchagin, E. E., Reshetnikova, Y. S., Golovina, N. I., & Brynza, N. S. (2020). Colour solutions in navigation support as an indicator of visualisation efficiency in a modern hospital (a review). *Kuban Scientific Medical Bulletin*, *27*(5), 128–143. https://doi.org/10.25207/1608-6228-2020-27-5-128–143

Lachance, J., Vinit, F., & Hillion, J. (2018). Exploration of touch in faculty of medicine? Presentation and exploration of touch and perspective in practice. *International Journal of Whole Person Care*, *5*(1). https://doi.org/10.26443/ijwpc.v5i1.160

Levy, A. G., Maselko, J., Bauer, M., Richman, L., & Kubzansky, L. (2007). Why do people with an anxiety disorder utilize more nonmental health care than those without? *Health Psychology*, *26*(5), 545–553. https://doi.org/10.1037/0278-6133.26.5.545

Lewis, A. O., & Nin, A. (1962). Seduction of the minotaur. *Books Abroad*, *36*(3), 322. https://doi.org/10.2307/40116995

Lewitz, J. (2008). Effective sound masking for speech privacy in open plan offices. *The Journal of the Acoustical Society of America*, *123*(5), 2971–2971. https://doi.org/10.1121/1.2932456

Littleton, H. L., & Grills-Taquechel, A. (2011). Evaluation of an information processing model following sexual assault. *Psychological Trauma: Theory, Research, Practice and Policy*, *3*(4), 421–429. https://doi.org/10.1037/a0021381

López-Cabarcos, M. Á., López-Carballeira, A., & Ferro-Soto, C. (2019). The role of emotional exhaustion among public healthcare professionals. *Journal of Health Organization and Management*, *33*(6), 649–655. https://doi.org/10.1108/jhom-04-2019-0091

Machingura, T., Kaur, G., Lloyd, C., Mickan, S., Shum, D., Rathbone, E., & Green, H. (2019). An exploration of sensory processing patterns and their association with demographic factors in healthy adults. *Irish Journal of Occupational Therapy*, *48*(1), 3–16. https://doi.org/10.1108/ijot-12-2018-0025

Majd, N. F., Golembiewski, J., & Tarkashvand, A. (2020). The psychiatric facility: How patients with schizophrenia respond to place. *Design for Health*, *4*(3), 384–406. https://doi.org/10.1080/24735132.2020.1846849

Marsh, J. E., Ljung, R., Nöstl, A., Threadgold, E., & Campbell, T. A. (2015). Failing to get the gist of what's being said: Background noise impairs higher-order cognitive processing. *Frontiers in Psychology*, *6*, 548. https://doi.org/10.3389/fpsyg.2015.00548

Marzban, S., Candido, C., Mackey, M., Engelen, L., Zhang, F., & Tjondronegoro, D. (2022). A review of research in activity-based working over the last ten years: Lessons for the post-COVID workplace. *Journal of Facilities Management*, *21*(3), 313–333. https://doi.org/10.1108/jfm-08-2021-0081

Matthew, A., Moffitt, C., Huth-Bocks, A., Ronis, S., Gabriel, M., & Burkhart, K. (2022). Establishing trauma-informed primary care: Qualitative guidance from patients and staff in an urban healthcare clinic. *Children*, *9*(5), 616. https://doi.org/10.3390/children9050616

McCleane, G. (1999). Pain symptoms and effects. *Practice Nursing*, *10*(13), 20–22. https://doi.org/10.12968/pnur.1999.10.13.20

McGreevy, S., & Boland, P. (2020). Sensory-based interventions with adult and adolescent trauma survivors. *Irish Journal of Occupational Therapy*, *48*(1), 31–54. https://doi.org/10.1108/ijot-10-2019-0014

McLachlan, F., & Leng, X. (2021). Colour here, there, and in-between – placemaking and wayfinding in mental health environments. *Color Research & Application*, *46*(1), 125–139. https://doi.org/10.1002/col.22570

Meredith, P. J., Andrews, N. E., Thackeray, J., Bowen, S., Poll, C., & Strong, J. (2021). Can Sensory- and attachment-informed approaches modify the perception of pain? An experimental study. *Pain Research and Management*, *2021*, 5527261. https://doi.org/10.1155/2021/5527261

Meredith, P. J., Bailey, K. J., Strong, J., & Rappel, G. (2015a). Adult attachment, sensory processing, and distress in healthy adults. *The American Journal of Occupational Therapy*, *70*(1), 7001250010p1–7001250010p8. https://doi.org/10.5014/ajot.2016.017376

Meredith, P. J., Rappel, G., Strong, J., & Bailey, K. J. (2015b). Sensory sensitivity and strategies for coping with pain. *The American Journal of Occupational Therapy*, *69*(4), 6904240010p1–6904240010p10. https://doi.org/10.5014/ajot.2015.014621

Metz, A. E., Boling, D., DeVore, A., Holladay, H., Liao, J. F., & Vlutch, K. V. (2019). Dunn's model of sensory processing: An investigation of the axes of the four-quadrant model in healthy adults. *Brain Sciences*, *9*(2), 35. https://doi.org/10.3390/brainsci9020035

Miller, L. J., Anzalone, M. E., Lane, S. J., Cermak, S. A., & Osten, E. T. (2007). Concept evolution in sensory integration: A proposed nosology for diagnosis. *The American Journal of Occupational Therapy: Official Publication of the American Occupational Therapy Association*, *61*(2), 135–140. https://doi.org/10.5014/ajot.61.2.135

Neumann, R. (2003). *Ten steps to achieve acoustic comfort in your office*. Ruth Neumann Architect. https://ruthnewman.com.au/ten-steps-to-achieve-acoustic-comfort-in-your-office/

Nightingale, F. (1859). *Notes on nursing: What it is, and what it is not*. Blackie & Son.

Noël, G. (2017). Health design: Mapping current situations, envisioning next steps. *The Design Journal*, *20*(Suppl 1), S2304–S2314. https://doi.org/10.1080/14606925.2017.1352746

Padmakumar, A. D., Cohen, O., Churton, A., Groves, J. B., Mitchell, D. A., & Brennan, P. A. (2016). Effect of noise on tasks in operating theatres: A survey of the perceptions of healthcare staff. *The British Journal of Oral & Maxillofacial Surgery*, *55*(2), 164–167. https://doi.org/10.1016/j.bjoms.2016.10.011

Park, J.-H. (2022). Introduction of sound healing as an energy therapy and implication of applying sound healing in psychotherapy. *The Association of Korea Counseling Psychology Education Welfare*, *9*(3), 27–46. https://doi.org/10.20496/cpew.2022.9.3.27

Pletschko, T., Felnhofer, A., Schwarzinger, A., Weiler, L., Slavc, I., & Leiss, U. (2016). Applying the international classification of functioning – children and youth version to pediatric neuro-oncology. *Journal of Child Neurology*, *32*(1), 23–28. https://doi.org/10.1177/0883073816669647

Postma, E. M., De Vries, Y. C., & Boesveldt, S. (2017). Tasty food for cancer patients: The impact of smell and taste alterations on eating behaviour. *Nederlands Tijdschrift Voor Geneeskunde*, *160*, D748.

Putrino, D., Ripp, J., Herrera, J. E., Cortes, M., Kellner, C., Rizk, D., & Dams-O'Connor, K. (2020). Multisensory, nature-inspired recharge rooms yield short-term reductions in perceived stress among frontline healthcare workers. *Frontiers in Psychology*, *11*, 560833. https://doi.org/10.3389/fpsyg.2020.560833

Queensland, Work Health and Safety. (2022). *Managing the risk of psychosocial hazards at work: Code of practice (QLD)*. Workplace Health and Safety Queensland. www.worksafe.qld.gov.au/__data/assets/pdf_file/0025/104857/managing-the-risk-of-psychosocial-hazards-at-work-code-of-practice.pdf

Robinson, J. (2001). Post-traumatic stress disorder: Avoidance behaviour. *British Journal of Midwifery*, *9*(12), 775–775. https://doi.org/10.12968/bjom.2001.9.12.9393

Ruotsalainen, J. H., Verbeek, J. H., Mariné, A., & Serra, C. (2016). Preventing occupational stress in healthcare workers. *Sao Paulo Medical Journal*, *134*(1), 92. https://doi.org/10.1590/1516-3180.20161341t1

Sakurama, M., Ito, M., Nakanowataru, Y., & Kooriyama, T. (2023). Selection of appropriate dogs to be therapy dogs using the C-BARQ. *Animals: An Open Access Journal from MDPI*, *13*(5), 834. https://doi.org/10.3390/ani13050834

Satoki, I., Eriko, T., & Masahiko, K. (2020). Effective medical creation (EMC) – a new approach to improvement of patient management in the standpoint of hospital room environment. *Open Journal of Anesthesiology*, *10*(12), 409–421. https://doi.org/10.4236/ojanes.2020.1012036

Scanlan, J. N., & Novak, T. (2015). Sensory approaches in mental health: A scoping review. *Australian Occupational Therapy Journal*, *62*(5), 277–285. https://doi.org/10.1111/1440-1630.12224

Schiffman, S. S. (2018). Influence of medications on taste and smell. *World Journal of Otorhinolaryngology – Head and Neck Surgery*, *4*(1), 84–91. https://doi.org/10.1016/j.wjorl.2018.02.005

Schmidt, N., Gerber, S. M., Zante, B., Gawliczek, T., Chesham, A., Gutbrod, K., Müri, R. M., Nef, T., Schefold, J. C., & Jeitziner, M.-M. (2020). Effects of intensive care unit ambient sounds on healthcare professionals: Results of an online survey and noise exposure in an experimental setting. *Intensive Care Medicine Experimental*, *8*(1), 34. https://doi.org/10.1186/s40635-020-00321-3

Sinha, Y., Silove, N., Hayen, A., & Williams, K. (2011). Auditory integration training and other sound therapies for autism spectrum disorders (ASD). *The Cochrane Database of Systematic Reviews*, *12*, CD003681. https://doi.org/10.1002/14651858.cd003681.pub3

Skeath, P., Jenkins, M. A., McCullough, A., Fine, A. H., & Berger, A. (2015). Increasing the effectiveness of palliative care through integrative modalities. In *Handbook on animal-assisted therapy* (pp. 261–277). Elsevier. https://doi.org/10.1016/B978-0-12-801292-5.00019-5

Solet, J. M., & Barach, P. R. (2012). Managing alarm fatigue in cardiac care. *Progress in Pediatric Cardiology*, *33*(1), 85–90. https://doi.org/10.1016/j.ppedcard.2011.12.014

Sorensen, G., McLellan, D. L., Dennerlein, J. T., Nagler, E. M., Sabbath, E. L., Pronk, N. P., & Wagner, G. R. (2019). A conceptual model for guiding integrated interventions and research: Pathways through the conditions of work. In H. L. Hudson, J. A. S. Nigam, S. L. Sauter, L. C. Chosewood, A. L. Schill, & J. Howard (Eds.), *Total worker health* (pp. 91–106). American Psychological Association.

Sorin, R., Brooks, T., & Lloyd, J. (2015). The impact of the classroom canines program on children's reading, social/emotional skills, and motivation to attend school. *The International Journal of Literacies*, *22*(2), 23–35. https://doi.org/10.18848/2327-0136/cgp/v22i02/48840

Stewart, L. A., Chang, C. Y., & Rice, R. (2013). Emergent theory and model of practice in animal-assisted therapy in counseling. *Journal of Creativity in Mental Health*, *8*(4), 329–348. https://doi.org/10.1080/15401383.2013.844657

Tanja-Dijkstra, K., & Pieterse, M. E. (2011). The psychological effects of the physical healthcare environment on healthcare personnel. *The Cochrane Database of Systematic Reviews*, *1*, CD006210. https://doi.org/10.1002/14651858.cd006210.pub3

Tokunaga, A., Akiyama, T., Miyamura, T., Honda, S., Nakane, H., Iwanaga, R., & Tanaka, G. (2019). Neonatal behavior and social behavior and sensory issues in 18-month toddlers. *Pediatrics International*, *61*(12), 1202–1209. https://doi.org/10.1111/ped.14033

Tumlinson, K., Gichane, M. W., Curtis, S. L., & LeMasters, K. (2019). Understanding healthcare provider absenteeism in Kenya: A qualitative analysis. *BMC Health Services Research*, *19*(1), 660. https://doi.org/10.1186/s12913-019-4435-0

Turjeman-Levi, Y., & Kluger, A. N. (2022). Sensory-processing sensitivity versus the sensory-processing theory: Convergence and divergence. *Frontiers in Psychology*, *13*, 1010836. https://doi.org/10.3389/fpsyg.2022.1010836

Wallin, L., & Eriksson, M. (2009). Newborn individual development care and assessment program (NIDCAP): A systematic review of the literature. *Worldviews on Evidence-Based Nursing*, *6*(2), 54–69. https://doi.org/10.1111/j.1741-6787.2009.00150.x

Wang, H., Tang, D., Wu, Y., Zhou, L., & Sun, S. (2020). The state of the art of sound therapy for subjective tinnitus in adults. *Therapeutic Advances in Chronic Disease*, *11*, 2040622320956426. https://doi.org/10.1177/2040622320956426

Warner, E., Koomar, J., Lary, B., & Cook, A. (2013). Can the body change the score? Application of sensory modulation principles in the treatment of traumatized adolescents in residential settings. *Journal of Family Violence*, *28*(7), 729–738. https://doi.org/10.1007/s10896-013-9535-8

Weaver, M. S., Robinson, J., & Wichman, C. (2020). Aromatherapy improves nausea, pain, and mood for patients receiving pediatric palliative care symptom-based consults: A pilot design trial. *Palliative and Supportive Care*, *18*(2), 158–163. https://doi.org/10.1017/s1478951519000555

Winter, A. L., Henecke, S., Lundström, J. N., & Thunell, E. (2023). Impairment of quality of life due to COVID-19-induced long-term olfactory dysfunction. *Frontiers in Psychology*, *14*, 1165911. https://doi.org/10.3389/fpsyg.2023.1165911

Work Health and Safety Act 2011, Pub. L. No. 137, C2018C00293. (2011).

Wright, L., Bennett, S., & Meredith, P. (2020). "Why didn't you just give them PRN?": A qualitative study investigating the factors influencing implementation of sensory modulation approaches in inpatient mental health units. *International Journal of Mental Health Nursing*, *29*(4), 608–621. https://doi.org/10.1111/inm.12693

Yap, S. F. (Crystal), Phillips, M., Hwang, E., & Xu, Y. (2022). Transforming healthcare service environments: A sensory-based approach. *Journal of Service Theory and Practice*, *32*(5), 673–700. https://doi.org/10.1108/jstp-02-2022-0033

Zaghloul, H., Pallayova, M., Al-Nuaimi, O., Hovis, K. R., & Taheri, S. (2017). Association between diabetes mellitus and olfactory dysfunction: Current perspectives and future directions. *Diabetic Medicine: A Journal of the British Diabetic Association*, *35*(1), 41–52. https://doi.org/10.1111/dme.13542

16 Integrating Human Factors and Ergonomics into the Design of Community Diagnostic Centres

A Journey towards Insight

Julie Combes and Mark Sujan

16.1 JULIE REFLECTS ON THE STATE OF DIAGNOSTIC SERVICES AND THE NATIONAL DIAGNOSTIC WORKFORCE CRISIS

The National Health Service (NHS) has been viewed by many in the United Kingdom as a 'national treasure' with access to infinite universal healthcare since its inception in 1948. However, the impact of COVID-19 has affected the capacity of the NHS, and we have witnessed a change in public experience and perceptions. Healthcare systems are struggling to keep up with activity: we are enduring a time of increased demand with population demographic and epidemiological transition. There are societal and political changes placing additional demands on an already constrained healthcare workforce. The NHS appears unsustainable, with a political overhaul deemed necessary. There is radical reform underway to drive improvements in population health and transform patient services to make sure that the NHS is fit for the future.

The elective care pathway and, more specifically, diagnostic services had been recognised pre-pandemic as needing radical investment and reform. The independent Richards Review 2020 (Richards, 2020) further outlined key recommendations for recovery and renewal of diagnostic pathways to manage backlog and implement beneficial change such as new models of community care. A key component of the new service models was the introduction of community diagnostic centres (CDCs) to provide rapid elective diagnostic services. These were intended to relieve pressure on acute sites and bring services closer to the patient. The CDC location could be within established NHS buildings or newly designed or built capital within the heart of communities such as shopping centres. Alongside this, new workforce models would be required, involving skill-mix initiatives, new roles, and advanced practice.

DOI: 10.1201/9781032711195-19

It would involve integrating more unregistered staff, learning from task (re)distribution, and the management of flexible roles during the pandemic.

However, the country is experiencing a workforce crisis that has not been seen on this scale before. Workforce projections suggest that an extra 475,000 jobs will be required in health by the beginning of the next decade. Not surprisingly, despite the need to widen access and to reduce waiting lists through the establishment of CDCs, there is widespread concern that workforce shortages are having an impact on the sector's ability to provide quality care and meet national performance targets.

16.2 JULIE TAKES ON THE WORKFORCE CHALLENGE

In June 2021, I was asked to meet with the London region's NHS England Diagnostic Programme Director and the Health Education England's Director of Performance. They were in the early phases of developing a comprehensive plan to implement the recommendations of the Richards Review 2020 and, more specifically, to establish the CDCs across several speciality care pathways, such as cardiac. The purpose of the meeting was to learn from the work that I had led during the COVID-19 pandemic and to understand the potential to apply any learnings to this workforce challenge.

During the COVID-19 first wave (March 2020–May 2020), I was responsible for leading the education and training workstream for NHS Nightingale hospital housed in ExCeL London, the first field hospital built in response to the projected increase in demand for critical care. This unique challenge meant that we had to design a workforce model to deliver care during a surge in demand. There were diverse roles and new ways of working within the context of a diluted skill mix due to a lack of available qualified staff. As we designed the workforce model, we were required to adapt and create a rapid training programme. It had to equip people with the fundamental skills, knowledge, and capabilities to deliver safe care while working as part of an unfamiliar team and in what some would consider a hostile working environment. It could be considered hostile due to it being a physical environment that has been transformed from a conference centre to a 'ward' with capacity of up to 2,000 intensive care beds in comparison to the usual 10–20 bedded ward. The lighting, equipment, acoustics, team structures and working practices were all foreign to the workforce, within the context of an emerging national emergency and social isolation.

Within this first meeting, it was clear that workforce shortages in some professions were a limiting factor to the expansion of CDCs and London's capacity to deliver optimum care and reduce the backlog. Initial workforce modelling for diagnostic services demonstrated that if the current clinical and workforce model were utilised, then supply would not meet demand, impacting patient and workforce outcomes. We needed to think differently to resolve the issues.

It was clear that there was a focus on rapidly increasing the integration of unregistered staff into the workforce, diversifying the skill mix, and freeing registered and uniquely qualified staff to undertake the higher order tasks. This was considered a 'quick fix' because it could ensure a quick supply of staff. The staff group needed to be rapidly trained utilising competency-based learning to increase the capacity and activity of CDCs.

16.3 MARK REFLECTS ON THE ROLE OF HUMAN FACTORS/ERGONOMICS DURING COVID-19

In March 2020, I received an early morning text message. It was from the CEO of the Chartered Institute of Ergonomics and Human Factors. He was looking for volunteers to develop a guidance document, within 48 hours, on human factors in the design of rapidly manufactured ventilator systems. I responded.

This was the start of a quite amazing and hugely rewarding journey in times of a national crisis. Many professions and professionals were taken by surprise by the pandemic: many of us did not know what to do, or how to react. Industries such as commercial aviation came to a standstill and many people across diverse sectors had to go on furlough.

The development of the guidance material suddenly provided the opportunity to do something, to contribute in some small way. The UK government had called for new manufacturers to voluntarily increase the scale of medical ventilator systems production in response to the expected demand for this vital life-saving medical technology. However, we recognised that there were design, usability, and safety challenges, especially for manufacturers unfamiliar with medical device design, that required further guidance. We developed the guidance document with the support of many volunteers and presented this to the regulator and government officials. Fortunately, the anticipated demand for additional ventilator capacity did not materialise. However, the guidance document (Sujan & Rashid, 2020) stands as a testimony to the role of HF/E, and it triggered a flurry of activity.

In collaboration with colleagues, CIEHF continued to develop HF/E guidance documents and held webinars and dissemination events throughout the pandemic. Based on the success of the ventilator guidance material, we developed a rapid usability testing protocol for ventilators. This was followed by guidance on how to design better work procedures, and – as we slowly emerged from the pandemic – guidance on how to capture lessons from COVID-19, and guidance on safe return to work. These activities have supported the NHS, and they have undoubtedly raised the profile of HF/E in the health sector. On a personal level, this has been extremely rewarding as an opportunity to contribute something while applauding those delivering care at the frontline.

16.4 JULIE ADVOCATES FOR A SYSTEMS APPROACH

I had some apprehension about applying a model developed during a national emergency and I was hesitant to promote or advocate the strategies when we were looking at creating a new model that would become 'business as usual'. This was an opportunity to share the experiences I had at NHS Nightingale. I could introduce some of the challenges that we experienced when we considered the workforce in isolation, rather than recognising the interconnectedness and interdependencies that exist within the complex NHS systems. For example, designing a new skill mix and training people to work a new way does not mean that organisations or teams will culturally or professionally accept and embed this into their practice. Organisations may not review and adapt their current workforce models or provide a budget to train and recruit into new posts. This could leave some of the workforce feeling frustrated, undervalued, or excluded because they are not given the opportunity to develop and apply new skills or recognition for the enhanced experience and skills they bring.

Through our discussions, I suggested that this was an opportunity to take a different approach, one that recognised the complexity of the system and utilised the scientific discipline of Human Factors and Ergonomics to understand interactions among humans and elements of the system. I shared my concerns that we continued to identify the same 'solutions' that, in most cases, only provide slight improvement. Rather, I advocated for an understanding and articulation of the underlying issues that needed to be resolved for any real benefit to be realised for patients and staff. I recommended that we adopt a human-centred approach, to 'diagnose' the genuine issues from the perspective of those interacting with services and delivering care. We needed to identify and co-create high impact actions and implement these to improve overall system performance and staff well-being.

16.5 JULIE EMBARKS ON A JOURNEY

The acceptance of the recommendation was received with full support immediately. In hindsight, I was not entering the meeting to propose or recommend any action but to simply share experience and learnings. In fact, by the end of the meeting, I was asked if I could lead this piece of work and pull together a proposal for the board to consider with associated budget. However, I am not a human factors or ergonomics specialist but would call myself a voyeur and advocate for the approach, having experienced the impact when applied in practice. I am not and did not feel qualified to lead this work, nor did I have the credibility to assure key stakeholders of the approach. I suggested that I contact the Chartered Institute of Ergonomics and Human Factors (CIEHF) to work in partnership with their specialty advisors and draw on the appropriate expertise.

I recognise that networks and relationships have been key to much of my recent experience during the pandemic and this also applied when reaching out to the CIEHF. I did not have any direct links, so I contacted a colleague with whom I had worked with five years previously. This colleague kindly introduced me to the CEO to discuss the high-level requirements. At the time, NHS England and Health Education England articulated a sense of urgency and set out a compressed period for proposed work to start and for outcomes to be realised. On reflection, this was also in line with the intense interest, political concerns, and publicity about patient outcomes and the growing waiting lists in diagnostic pathways. I felt a sense of responsibility and with this came some anxiety with the reality that this was a complex piece of work with multiple interdependencies that would be out of our control, impacting what we could achieve and potential benefits from this approach. I felt colleagues' expectations that this proposal needed to resolve workforce issues in diagnostics and solutions had to be articulated quickly and in a prescriptive manner such as numbers and activity relating to performance. It was clear that public and political anxiety were putting intense pressure on NHS leaders in the London region, with multiple directives coming from the government to provide assurance and plans for recovery. However, drawing on the experience I had during the first wave of COVID-19, I recognised the demands and behaviours from NHS leadership and governmental ministers and was able to navigate this personally and professionally with a bit more ease. I like to think of Rudyard Kipling's words from the poem 'If' and would draw on this on many occasions before responding to the latest requests.

> If you can keep your head when all about you
> Are losing theirs and blaming it on you,
> If you can trust yourself when all men doubt you,
> But make allowance for their doubting too.

All too often I have found that during a crisis, decisions can be made in haste and actions occur in response to pressure generated by fear. Action can be seen, it is what is expected, and by doing 'something' (read: 'anything') we may feel a level of assurance and comfort, even if it is not what is required and does not resolve the crisis overall. However, it is my experience that taking a little time to gather all the data available from various sources and considering interconnections, decision-making can be better informed and, therefore, more likely to lead to meaningful and sustained improvement. In some ways, I was drawing from the human factors training that I had as a clinician. I drew from my learnings about the importance of communication, team contributions, maintaining situational awareness, and recognising what impacts personal cognition and, therefore, response and performance.

At this time, I also knew that having a strong and professionally relevant multidisciplinary team to undertake this work was imperative. It was a programme of work that required a diverse skill set with differing levels of influence, technical expertise, and leverage within the sociopolitical landscape. The multidisciplinary team included HF/E specialists, workforce director, policy leads, clinicians, clinical director, programme management, and others were drawn in during the life cycle of the project as required. In working within this context previously, the credibility and interpersonal relationships of the multidisciplinary team have been pivotal to success, maintaining open lines of communication and promoting a flexible and respectful environment to engage in.

16.6 MARK DRAWS A MAP WITH JULIE

The CIEHF response to COVID-19 had been strong and successful and decision-makers started to notice the discipline. On the back of this, many new relationships with people across the NHS were formed. As a result, CIEHF were approached by Health Education England and NHS England to submit a proposal for the support of the design of CDCs.

Within CIEHF, we rapidly scoped the task and put together an expert team of Chartered Ergonomists. We developed an initial proposal based on a systems approach with a strong focus on organisational learning. This proposal was the starting point for discussions with stakeholders, and I was delighted to find in Julie a competent and knowledgeable ally.

In collaboration with our colleagues, Julie and I mapped out the high-level requirements for adopting a systems approach to the design of CDCs and for evidencing the role of HF/E. Julie was insistent that the siloed thinking in terms of clinical pathway design, on the one hand, and of workforce development, on the other hand, would need to be overcome and cojoined.

Eventually, we agreed that we would develop an accessible and evidence-based roadmap for stakeholders which sets out HF/E principles to consider in the design of CDCs. At the end of the project, there were ten principles focusing on identifying opportunities for system change, improving the interactions between different elements of the work system, and learning and adaption (Figure 16.1).

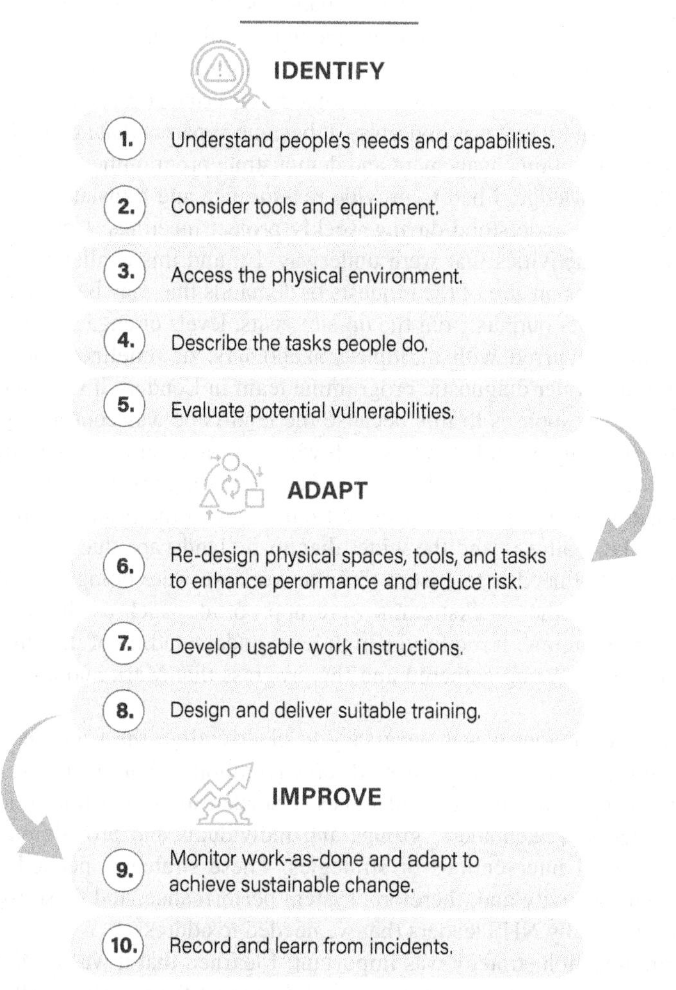

HEALTHCARE INSIGHTS:
VOICES OF THE CONSUMER, THE PRACTITIONER, AND THE WORK DESIGN STRATEGIST
IDENTIFY
1. Understand people's needs and capabilities.
2. Consider tools and equipment.
3. Access the physical environment.
4. Describe the tasks people do.
5. Evaluate potential vulnerabilities.
ADAPT
6. Re-design physical spaces, tools, and tasks to enhance perormance and reduce risk.
7. Develop usable work instructions.
8. Design and deliver suitable training.
IMPROVE
9. Monitor work-as-done and adapt to achieve sustainable change.
10. Record and learn from incidents.

FIGURE 16.1 Ten human factors/ergonomics principles for the design of CDCs.

Source: Adapted with permission from 'Vaccinating a Nation' (CIEHF, 2021).

16.7 JULIE BUILDS BRIDGES

On reflection, this is where it became clear that there was a difference in language, philosophical frameworks, familiar approaches, and the meaning associated with language among NHS England, my colleagues, Mark, and me. Many at NHS England had never heard of 'ergonomics' other than the design of equipment and envisaged Human Factors as team training. This highlighted to me the importance of communication, removing jargon from all types of communication that could be exclusionary and making no assumptions of levels of experience or mental models. Unfortunately, in initial conversations I shared the theoretical model to be used, the 'Systems Engineering Initiative for Patient Safety' (Holden & Carayon, 2021), and this caused some misperception. NHS England colleagues understood this to mean that patient safety was the dominant focus. I was advised that we needed to use the term 'workforce' in any communications with stakeholders, and that 'patient safety' was being used excessively, which could lead to misunderstandings among key influential groups. Concerns were raised that 'workforce' and 'optimising system performance' needed to be better articulated when presenting to stakeholders so that the described outcomes could be clearly aligned to the activity that was underway. It became apparent that agreed terminology was needed to maintain engagement and demonstrate programme relevance.

With this knowledge, I had to provide reassurance and translate the language to ensure that it was understood during weekly project meetings. I provided tangible examples of the activities that were underway. I found this challenging because in each meeting I was unsure of the requests or demands that may be made. It could be about different data outputs from the on-site visits, levels of engagement or involvement that had occurred with multiple stakeholders, or timelines and content for reporting to the wider diagnostic programme team in London. It was challenging to provide concise responses to this because the landscape was continually changing, with clinicians and key stakeholders with whom we needed to work pulled by their competing demands and clinical activity often overriding this project.

I also had to ensure that I kept myself current with the priorities in diagnostics within the CDC pathway and the wider diagnostic landscape due to potential interdependencies. We needed to ensure that the work we were doing was relevant and interactions with other workstreams were appropriate, such as the digital system integration programme. It required a high-level understanding of the emerging findings from the data being gathered and the progress that Mark and colleagues were making.

It became clear that it was necessary to clearly align the priorities articulated within diagnostic services, such as workforce retention and productivity, to data that was gathered by Mark and his colleagues. I needed to demonstrate relevance and credibility across stakeholders' groups and individuals and provide early insights into the potential interventions or strategies. These strategies needed to improve workforce productivity and, therefore, system performance, and these were the priorities articulated by NHS leaders that we needed to address.

A communication strategy was important: I learned that it was important to use regular and custom communication strategies to meet the diverse needs of multiple stakeholder groups at various levels of the system and during different stages of the project. During the initial stages of the project, a high-level stakeholder analysis was undertaken to identify those that could inform, influence, or be impacted by any change. Stakeholder groups included professional bodies, NHS trusts, policy groups, clinical leads, community representatives, or general practice groups. By scanning the landscape with colleagues from NHS England, we identified existing groups, boards, or networks with whom we could engage.

We understood that unique messaging (modes and register of communication) to meet the needs of these groups would be required. Therefore, we utilised email, attendance at key meetings and workshops to gain wider sense of checking, consensus and intelligence on initial data synthesis, and some one-to-one conversations with individuals who would further inform or disseminate information.

However, we also needed to engage and involve the clinical teams to understand work-as-done (the realities of work) and ensure that the initial findings reflected reality (the findings required validation). Also, clinical team involvement ensured better buy-in and project engagement. Emails were sent to solicit these responses and leaflets were provided. This did not gain the level of response or interaction that we required because many clinicians did not have working capacity to regularly check emails. We had to revisit the methods and draw on personal relationships and in-person interactions when on-site with the clinicians to involve them.

16.8 JULIE SETS A COURSE

I felt the pressure to quickly generate and report on outcomes that could assure the stakeholders of our performance and start to identify where support for clinical teams could be focused to improve their working conditions. Nonetheless, at times, due to the multiple competing demands of colleagues in NHS England and stakeholders within clinical practice, we experienced delays. I negotiated with NHS England diagnostic programme leads to ensure that the approach that we had agreed was given the necessary resource and time. We needed to involve the clinical leads and key stakeholders, such as professional bodies. The voice of the workforce was critical. However, this introduced some risk to the London diagnostic team because this meant utilising an approach that was outside of the conventional recruitment, retention, and training strategy undertaken in most regions or workforce programmes.

The desire to move quickly and find that 'silver bullet' to resolve the workforce crisis and improve productivity is persistent, but at least now there is some recognition that we need to consider the whole sociotechnical system to improve healthcare services and staff well-being. From my experience, the discipline of HF/E is most suited to the challenge.

In undertaking a systems approach, we uncover the performance limiting factors that existed among work systems and which influenced processes and outcomes. There was a presumption that introducing another role such as the healthcare support worker (un-registered staff) could relieve the Health Care Scientist (HCS) of some tasks during echocardiography, freeing them to see more patients and reduce backlog. However, when undertaking task analysis, it became clear that there were limited tasks that a healthcare support worker could undertake, and that supervision of unregistered staff could increase workload and reduce productivity of the HCS. Wider work system analysis was able to uncover multiple performance-limiting factors which also caused staff frustration, such as a lack of digitally enabled interoperable systems; poor communication across services; a lack of administrative support; absence of prospects for career progression; lack of supervision; and workload. It was clear that isolated interventions were unlikely to lead to sustainable improvement in productivity or staff satisfaction and their intention to stay. Such data allowed us to present an alternative lens to understand performance and productivity, focusing on the design of systems, including the organisational context, the equipment, the job design, the tasks, the environment, and the people to support good performance and improve job satisfaction, which we considered was essential during a national workforce crisis.

Ten Human Factors and Ergonomic principles were derived from this case study to provide an accessible way for including systems thinking in the design of imaging and diagnostics workstreams. The ten HF/E principles are generic, in as far as they can be used to identify areas for action and improvement across different settings. However, the specific nature of improvement interventions will be dependent on the local context, and different improvement interventions will be generated for different settings. An interactive digital infographic has been developed to provide more information[1] and a summary of the principles is shown in Figure 16.1.

Reflecting on my role and experience as an intermediary and mediator of the integration of ergonomics and human factors in the design of CDCs, I understand that this requires patience, diplomacy, and, most importantly, tenacity. This occurs within the context of collaboration and cross-disciplinary partnership among organisations, again demonstrating to me the importance of a multidisciplinary team to support collective action. Due to an open and safe environment created by the team, diverse views and experiences were shared and valued. Ideas were tested and challenged, providing me with the space and time to think differently and grow a deeper appreciation and interest into the discipline of Human Factors and Ergonomics in healthcare.

16.9 THE JOURNEY CONTINUES

There is an increasing interest in HF/E and an appreciation of the contribution that the discipline can make to the design of health systems and the delivery of care (Sujan et al., 2021). While there are encouraging stories of successful collaborations, such as the integration of HF/E principles into the design of CDCs, we have a long way to go. We will need to enhance and professionalise HF/E knowledge among healthcare professionals with active roles in quality improvement and patient safety. For example, CIEHF launched a 'healthcare human factors learning pathway' to support both academic and learning-at-work routes to achieve the status of technical specialist or TechCIEHF (Healthcare). Healthcare organisations also need to provide opportunities and career pathways for embedded suitably qualified HF/E practitioners (e.g., Chartered Ergonomists). These embedded HF/E practitioners can

[1] https://cdc-humanfactors.netlify.app/

work alongside all levels and groups of staff, building relationships and contributing technical expertise. In this way, the organisations can apply HF/E consistently in the design of work to maximise patient and staff safety and well-being, and the overall system performance.

REFERENCES

Chartered Institute of Ergonomics and Human Factors (CIEHF). (2021). *Vaccinating a nation*. Retrieved August 15, 2023, from https://ergonomics.org.uk/resource/vaccinating-a-nation.html

Holden, R. J., & Carayon, P. (2021). SEIPS 101 and seven simple SEIPS tools. *BMJ Quality & Safety*, *30*, 901–910.

Richards, M. (2020). *Diagnostics: Recovery and renewal – report of the independent review of diagnostic services for NHS England.* NHS England. www.england.nhs.uk/wp-content/uploads/2020/11/diagnostics-recovery-and-renewal-independent-review-of-diagnostic-services-for-nhs-england-2.pdf

Sujan, M., Pickup, L., Bowie, P., Hignett, S., Ives, F., Vosper, H., & Rashid, N. (2021). The contribution of human factors and ergonomics to the design and delivery of safe future healthcare. *Future Healthcare Journal*, *8*(3), e574.

Sujan, M., & Rashid, N. (2020). *Human factors in the design and operation of ventilators for Covid-19*. Chartered Institute of Ergonomics and Human Factors (CIEHF). Retrieved August 15, 2023, from https://ergonomics.org.uk/resource/design-guidance-for-ventilators.html

17 Resolving Complexity and Rehumanising Healthcare through Design Partnerships

Satyan Chari and Evonne Miller

17.1 BACKGROUND

Despite a popular cliché, it's almost certain that the healing profession preceded all others in society. Modern hospital-based medicine has come a long way from the ancient healing arts, but there is a growing sense that the sector's relentless pursuit of professionalisation, advancement, and sophistication has come at the cost of what was a defining quality for most of its long history – that is, healthcare's identity as a uniquely personal, trust-based vocation. For the past 30 years, healthcare has prioritised a focus on achieving well-orchestrated and predictable interactions among its multitude of specialised components. While important, this single-mindedness may have unintentionally pushed consumers away from the very place that they should be – at the centre of our collective attention. Thankfully, through the advocacy of consumers, and a gradual realisation within the sector, the pendulum is starting to swing back towards the people who rely on our services.

Modern hospital-based medicine has come a long way from the ancient healing arts, but there is a growing sense that the sector's relentless pursuit of . . . advancement . . . has come at the cost of . . . healthcare's identity as a uniquely personal, trust-based vocation.

17.2 THE CHALLENGE OF CONSUMER-CENTRED HEALTHCARE IMPROVEMENT IN COMPLEX SYSTEMS

The last decade has seen an explosion of research, strategy documents, government programmes, and improvement initiatives, all calling for a person-centred paradigm. Phrases like 'co-design', 'coproduction', and 'customer experience design' have become commonplace terms in healthcare. We can also see global initiatives to grant influence and control of healthcare services to consumers through the creation of consumer advocacy bodies, integrating consumer involvement into quality metrics and more. Yet, a curious tension is emerging.

DOI: 10.1201/9781032711195-20

Mannion and Exworthy (2017) describe this tension as a clash of 'institutional logics' – the logic of customisation and that of standardisation. They borrow the metaphor of the Procrustean Bed to describe how this paradox typically plays out in healthcare. Drawn from Greek mythology, the story describes a sinister character, Procrustes, who would seek to entice weary travellers to break their journey on the long road and partake of a restful night of sleep on a magical bed which could resize itself to suit the user. However, Procrustes proceeded to shackle those that accepted his offer, stretching and dismembering their body to fit his bed which was not at all customisable. In the paper, we liken the tortures inflicted by Procrustes to the task of healthcare workers trying to support an ideal of a personalised form of healthcare in a system that, over many decades of redesign, has been configured on principles of standardisation and opposes such efforts.

A second barrier to consumer-centred service enhancement in healthcare is the issue of complexity. The rise of complexity in healthcare has received academic and practical attention in recent years (Kannampallil et al., 2011; Braithwaite et al., 2017). McRae and Stewart suggest that healthcare of the modern era 'is better understood as perhaps 20 different industries, many of which need to seamlessly interact at critical junctures throughout a patient's journey' (p. I1039). This 'networked' complexity of healthcare is intuitively recognisable by those working within the system and reflects the complex reality of hospital-based care. This complexity is one source of resistance to improvement efforts, especially when these efforts go against the natural flows and rhythms of everyday work. With so much interdependency and interplay among systems and processes, it can be challenging for consumers and clinicians to know where to act or how to make sustained changes even when obvious problems exist. The practical challenge is that 'improvement work' needs to happen in an environment in which there are competing priorities, demands, and expectations – options that are seen to be innovative and necessary by frontline workers and consumers might be viewed as unacceptably laden with risk or a low priority by others whose attentions are on different (equally pressing) considerations. In this chapter, we will discuss the unique role that designers can play in navigating the tensions of local innovation versus system risk avoidance and in crafting novel paths out of the improvement 'impasses' that clinician-consumer teams typically face.

17.3 CO-DESIGN AND DESIGN COLLABORATIONS: AN IMPORTANT DISTINCTION

The term 'co-design' is frequently used within healthcare and has come to represent an enhanced commitment to incorporating the perspectives and aspirations of healthcare consumers in service improvement. This has undeniably created a more conducive environment for healthcare consumers to have their voices heard in healthcare change efforts. Yet, managers develop co-design projects in healthcare without active involvement of true design expertise or consideration of what role design specialists might play in an initiative. Thus, the level of design capability that they bring to each healthcare co-design project can vary. While there may not be objective data on the relative benefit that comes from bringing designers into healthcare co-design projects, our experience is that when professional designers are involved, the results can be transformative.

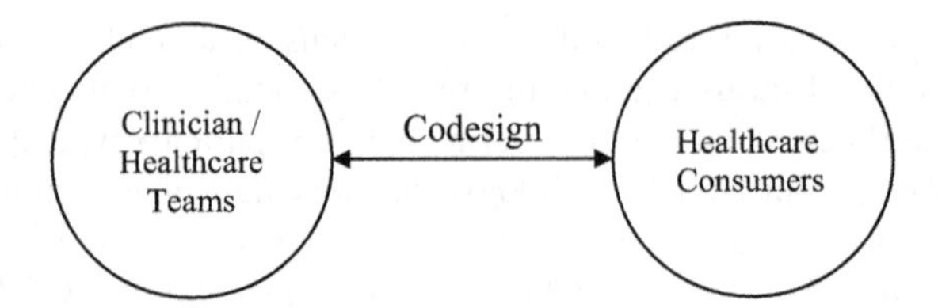

FIGURE 17.1 Prevailing view of co-design in healthcare.

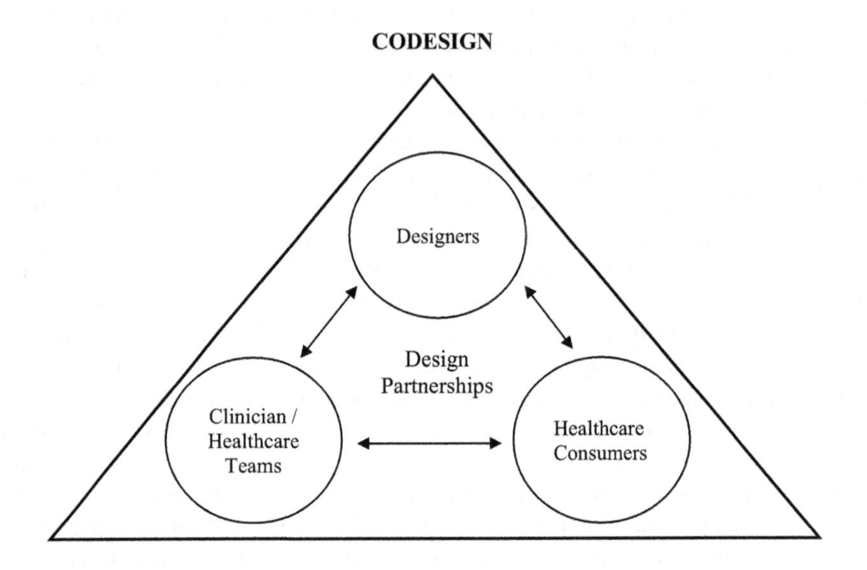

FIGURE 17.2 Codesign as an emergent feature of productive design partnerships.

As the goal of this chapter is to describe the genesis and effectiveness of a specific model where creative designers, clinicians, and consumers work in partnership, we have chosen to specify this distinction by using the phrase 'design collaborations' to describe the work we led. Within this framing, we view co-design as something that emerges (generatively) from productive design partnerships rather than something that is situated at the level of process. Figures 17.1 and 17.2 illustrate this distinction.

17.4 THE STORY OF DESIGN IN HEALTHCARE IN QUEENSLAND: THE GENESIS OF A RELATIONSHIP AND A MODEL

This chapter describes a collaborative relationship and a model that grew at the intersection of two communities, namely the creative design academic community from the Queensland University of Technology (QUT) and various parts of the publicly funded Queensland healthcare system. While there had been a long history of sporadic collaboration between academic designers and healthcare departments in Queensland, it was through the efforts of the QUT Design Lab and Clinical Excellence Queensland (part of the Queensland Department of Health) that work began in earnest in 2018 to create the networks and conditions to seed a sustained programme of collaboration.

At the early stages of this relationship, we foresaw many intersections and possibilities, but recognised the limited cross-disciplinary familiarity that existed between healthcare practitioners and design academics. This created a level of uncertainty about the possible types of collaborations, what meaningful collaboration might look like, and the practical challenges that might lay ahead. Thus, the focus of our work in these initial stages was on creating linkages and facilitating exploration. We undertook this in a pragmatic fashion – harnessing opportunities to bring the communities together, trying different strategies to catalyse partnered activities, and iterating frequently. When collaborative ideas picked up momentum, we sought to link those teams with funding and support to take those ideas further. We found the ideal opportunity to conduct our first 'safe-to-fail' learning experiment in 2019.

Each year, Clinical Excellence Queensland hosts a large in-person showcase event to promote leading innovations across the State and to inspire and connect Queensland Health staff passionate about healthcare improvement. The 2019 showcase received many innovative submissions (100+) that provided a practical starting point from which to identify teams that might have interest in collaborating with designers on the next phase of their work. Separate to the showcase selection process, a panel reviewed submissions with the intent of inviting a small number of teams to our initiative. The panel comprised of staff from the healthcare improvement unit (within CEQ) and the Director of the QUT Design Lab. We looked for initiatives that demonstrated a strong affinity for co-design and those that had the most potential for enhancement through creative design input rather than by focusing on any metric of 'maturity' or 'impact'. The panel selected eight diverse projects and invited the contributors to take part in the exploratory initiative. These projects ranged from person-centred palliative care to the consumer-led development of community maternity hubs and service redesign in the emergency setting.

Six project teams agreed to participate. These teams (clinicians, support staff, and consumer representatives who had worked with these teams) were requested to prepare a short 2-minute video 'backstory' describing their programme, conveying their passion for their work, and ending with where they would like to go next. Once the team produced their project videos, the QUT academic design community viewed these and 'bid' to work with individual projects that they found most interesting, or where there were clear convergences among a team's goals and the designer's areas of expertise. When the panel matched the appropriate designers with the healthcare teams, they encouraged them to connect over an online collaboration platform and start developing their pitch. All teams came together at the QUT Design Lab campus during the QUT 'Change by Design' Asia-Pacific Symposium in October 2019. Teams had 45 minutes together in a design sprint format to prepare a 3-minute pitch to win a six-month package of mentorship from the QUT Design Lab. The teams delivered their pitches at a special healthcare-focused session at the conference to a panel of peers and healthcare consumers.

This exercise was remarkably successful. We witnessed a cascade of collaborative activity from the pitch event, much of which would have been difficult to predict (or imagine) at the start. Many of the designer–clinician teams made rapid progress towards their pitched ideas, with new and innovative collaborations continuing to emerge after the event, as 'adjacent possibilities' were continuously discovered.

When we reflected on the outcomes of this initial 'seeding' activity, certain things became evident. We had introduced designers into established healthcare improvement project teams that came in with considerable experience in undertaking co-design activities with consumers. Yet, these engagements invariably led to novel ideas, important shifts in direction, and momentum-amplifying design activity. These activities included facilitation of creative processes and the development of various design artefacts, including visualisations, wireframes, and rapid iterations of low-fidelity prototypes. In short, designers participated far less as mediators of the co-design process, and far more as independent professionals contributing from a distinct paradigm of expertise, ideas, experiences, and networks, but in partnership with healthcare staff and healthcare consumers.

Another critical insight was that unlike the well-documented challenges that other fields have experienced in integrating their expertise into healthcare (Perry et al., 2021), we observed natural synergies (both conceptual and practical) when designers, healthcare teams, and consumers connected around improvement challenges. Before the event, we planned to support the project teams to help 'translate' idiosyncratic healthcare language, systems, and priorities into understandable constructs for designers and to manage any resistance arising from clinicians (with their experience in more linear, process-oriented forms of improvement) in participating in fluid, creative, and divergent design experiences. However, these concerns were unfounded. Designers reported that they were deeply drawn to the opportunity to create positive impact for the communities which they were a part of, and in working alongside healthcare providers and users on real-world care delivery issues. Designers were motivated by the passion exhibited by healthcare staff and consumers to address care needs, while clinicians and consumers reported renewed enthusiasm and excitement arising from their projects because of how the designers were able to navigate seemingly insurmountable issues with ease and bring entirely novel solutions to the table. Armed with these insights, we felt more confident in the face validity of the model, and we were ready to test at scale.

17.5 COVID-19 AND THE BIRTH OF THE HEALTHCARE EXCELLENCE ACCELERATOR (HEAL) BRIDGE LAB

Shortly after the completion of the pitch event in late 2019, the advent of the COVID-19 pandemic in the first quarter of 2020 confronted us. In an exceptionally tumultuous time, health systems in Australia set aside their ideas and priorities on long-term change and adopted a far nimbler approach to problem-solving and innovation. This unfamiliar environment created the conditions for the Queensland Health Bridge Labs programme to be born. Coupled with greater availability of internal resources to support rapid healthcare innovation, the disruptive effects of the pandemic on university-based teaching created a window of opportunity to draw in a larger number of academic designers to support the goals of the programme.

The Bridge Labs programme was conceived as an 'adhocratic' (flexible, organic, and decentralised) collaboration initiative, the focus and structure of which was shaped by insight gained through the earlier phase of work described. The linkage between the QUT Design Lab and Clinical Excellence Queensland was formalised as the Healthcare Excellence AcceLerator (HEAL) Bridge Lab and co-led by Professor Evonne Miller (QUT lead) and Dr Satyan Chari (CEQ Lead). A small internal team within Queensland Health administered the relationship (from developing flexible contracts and devising simple rules of engagement) and served as the primary enablers of cross-community linkages. The programme first sought to establish 'many-to-many' relationships and seeded a multitude of funded and in-kind collaborative projects spanning consumer-centred experience redesign in paediatric intensive care units, supply chain innovation in regional Queensland, consumer co-creation of a telehealth-based interdisciplinary care clinic, the production of design guidelines for safer and more comfortable personal protective equipment, child-centred redesign of paediatric emergency waiting spaces, and more. The variety of projects undertaken and the multitude of impactful outcomes generated have been previously described at length (Clinical Excellence Queensland, 2023; QUT Design Lab, 2023). In its third year of operation, we are now able to recognise two main ways in which design partnerships benefit healthcare. We describe these in the following sections.

17.6 THEME 1: OVERCOMING BARRIERS TO IMPROVEMENT

The challenge of creating impactful and lasting improvements in healthcare has been a consistent theme in research and practice for decades (Macrae & Stewart, 2019). Healthcare improvement is not known to be easy and is often characterised by a slow pace of progress or inertia despite efforts to overcome this. Part of the problem might well lie in the tools that healthcare has traditionally relied on to drive change. Political, procedural, and educational methods have their place, but these methods can worsen (rather than improve) issues when there is already a level of resistance to change. We discovered that in these situations, designers employed an entirely different set of strategies and skills to help teams move forward and by doing so, they did not elicit the same degree of 'pushback' when introducing novel ideas.

One reason appears to be that designers who participated in HEAL Bridge Labs projects proved to be adept at moving between the dual roles of 'lead' creative designer and design-thinking facilitator as the situation required. In practice, what we observed was that even in early stages of project familiarisation, designers would reframe information that they gathered into a variety of viewpoints and present it back to the teams for clarification and elaboration. This process could manifest in the form of clinicians describing a set of functional goals for a service, consumers describing their desired outcomes and pain points, and the designer then pulling this information together as an impactful illustration (Urine Contamination Project, QUT HEAL catalogue, 2023a), an empathy map (VOICED project overview, Cheers, 2023), user personas ('just' Healthcare for CALD clients, QUT HEAL catalogue, Cheers, 2023a), incorporating these various elements. The production of the design artefacts were inherently creative processes, but designers would deftly use these to facilitate subtle, but sometimes profound, shifts in the conversation, to develop empathy for a slightly different narrative, to generate a shared enthusiasm for a novel direction of progress and overcome perceived critical barriers to progress. This duality in the designer's process holds deep lessons for the global practice of healthcare improvement which has for many decades shied away from 'social-relation-creative' innovation activities, in favour of 'sequential-mechanistic' improvement approaches (with little success).

An illuminating example was a HEAL project to support the design and development of a virtual outpatient integration service which aimed to bring multiple medical specialists together to provide consumer-centred care for people with complex chronic conditions. The healthcare teams were deeply motivated by the potential for enhancement of the consumer experience and a multitude of healthcare efficiencies that this programme would generate. However, the programme (in its nascent state) had been conceptualised entirely in the language and logic of healthcare practitioners – describing the core functional components of the programme, conveyed in healthcare-centric illustration (not unlike Figure 17.2) showing three icons representing different medical specialties at the vertices of a triangle and with an icon representing the patient at the centre of the triangle. Whilst instantaneously understandable to healthcare professionals, the experience designer working on the project started her inquiries in a completely different (conceptual) space to focus of the clinical team. The team started to discuss questions: '*Who is the service for? What do we know about them? Where do they live? What is their typical experience of accessing healthcare?*' These questions helped ground a deeper and different understanding of proposed service users (Stroke mapping, Cheers, 2023a) which gradually moved the explorations of what the experience of a virtual outpatient clinic might entail (VOICED project overview, Cheers, 2023b), establishing the constraints of what might be possible (in terms of customisation and personalisation) from a consumer experience perspective. Subsequent thinking naturally gravitated towards questions as follows:

> Where would patients access the service (the patient's home or at a local community health centre)? What would the check-in procedure entail? What might be the unexpected issues to consider before implementing the new model (changing dynamics, anxiety from multiples clinicians in the same virtual room et cetera)?

These conversations yielded several designed artefacts (resources) that helped iteratively refocus and shape the conversation. However, they also became valuable catalysts for meaningful co-design with service consumers in the next stage of the project.

Another exemplar of design-led sensemaking was a HEAL project where a visual/experience designer was engaged to help a state-wide clinical network comprised of many different professions, health services covering diverse catchments (metropolitan, regional, rural, and remote regions of Queensland), and consumer groups. The goal of this project was to assist the group in finding ways to engage with big system-level issues concerning stroke care across the State and develop consensus on the types of changes to prioritise. A common issue that besets large complex priorities (with many stakeholders and diverse local contexts) is that it can be exceedingly difficult to describe the problem in a manner that incorporates a sufficient cross-section of perspectives without abstracting it to a point that renders it completely sterile and unactionable. The toolkits of project- and change-management are often deficient in managing this tension well (that is, in finding a central narrative while retaining important diversity of viewpoints) and typically move (all too) quickly to a crystallised definition of the problem and then to a solution.

Instead, the designer supporting the stroke clinical network began by developing several scenario-based visualisations of patient journeys based on her evolving understanding of the problem space. These visualisations rendered complex information on location-dependent response times, the contributions of various service providers, the pain-points, and bottlenecks. Members reflected on these visualisations for testing, validation, and elaboration. Through the process of collaboration with their design partner, co-shaping the information that went into these visualisations, the network discovered new insights and empathies for each other's sets of challenges and aspirations. What is most worth noting is that the iterative cycles of information capture, rendition, and reflection did not make the situation any less complex, but rather allowed network members to navigate this complexity in a more productive manner and help each other identify opportunities with the greatest leverage without resorting to early compromise or excluding perspectives and needs of various stakeholders along the way. This project clearly showcased the natural complementarity between creative and empathic sensemaking approaches and the navigation of complex, systemic, and multi-causal issues that characterise the safety, quality, and equity challenges in healthcare today.

These two exemplars of designer-led problem reframing shared common characteristics with other projects that we supported. First, the designer explored the problem with an inquisitive mindset. Designers did not seek to critique or impose, but naturally sought to elicit various perspectives with the aim of supporting understanding and discovery. This approach, which diverges markedly from the conventional (analytical) approach of healthcare 'improvement science', seemed to energise (rather than exhaust) teams by validating the complex lived realities of participants and recipients of healthcare service delivery. How designers naturally approach problems is through what David Cooperrider (Cooperrider et al., 2008) has termed appreciative inquiry (AI) – an approach which purposely moves away from the dominant problem-focused deficit discourse to a more strengths-based, generative, and constructive approach. AI encourages change agents to re-examine their organisation, systems, and people with 'appreciative eyes', rather than the traditional deficit-based model of problematising change and focusing on 'what's wrong, what's the problem' (Cooperrider, 2017). AI has been a paradigm shift in organisational problem solving, with the purposeful focus on strengths, achievements, and opportunities creating much needed space for a positive dialogue about the future and change. AI is, as Whitney et al. (2010) explain, 'the search for the best in people, their organisations, and the strengths-enriched world around them' (p. 1), and central to a broader focus on positive organisational development. While AI conceptually aligns with the design-thinking process because both encourage positive thinking and imagining about the future, the two approaches remain unlinked in most instances. Notably, in a recent expansion of the AI framework – into innovation-inspired positive organisation development (IPOD) – Cooperrider explicitly links design theory with appreciative inquiry, positive organisational scholarship, positive psychology, and design thinking (Cooperrider & Lindsey, 2011). This positive focus on 'what might be' is an essential component of the design approach – and central to the HEAL ethos.

Secondly, designers were skilful in using creative design modalities (through sketching, digital illustrations, infographics, and many other means) that allowed them to seamlessly fold beneficial ideas into conversations in ways that promote forward movement. The visual language of design enables the communication of complex and abstract ideas in an engaging, accessible, and memorable way. Drawing – whether it is sketching a self-portrait, a work experience, or initial prototype ideas – helped to create a shared understanding between diverse stakeholders (consumers, clinicians, designers etc.), and the integration of drawing activities into co-design/design thinking sprints also helped to create a more relaxed, enjoyable environment, which in turn fostered creativity, collaboration, and innovation.

Designers know that drawing is a great ice-breaker activity. Based on David Cooperrider's positive strengths-based approach of appreciative inquiry (Cooperrider, 2017), HEAL workshops often prompted participants to start by drawing a '*moment of exceptional practice when they felt inspired and motivated at work*'. This starts a workshop positively, and there is typically an excited air as people sketch and share work highlights. At one co-design workshop, for example, clinicians and consumers sketched and shared highlights of when a patient spoke for the first time in six weeks (saying 'I love you, Mum'), of taking a patient – after 14 months of in-patient treatment – to visit their rural property for a picnic, of a ward Christmas Party (bringing staff and patients together to celebrate), and when staff worked to bring a much beloved and missed pet dog into the hospital to visit its owner, as we see in Figure 17.3. The specific instructions were as follows:

> Recall a special moment of exceptional practice when you were really engaged, excited, and proud of your work. *Take 4 minutes to remember and draw this experience.* Add a title, key descriptors (dot points), and your name (optional).
>
> *Table Share (3 minutes):* Share your exceptional experience with the group: *What were the common themes?* Pick one story that illustrates the shared themes.
>
> *Joint Analysis (3 minutes):* Each table shares one story with the other groups. As a table responds to the question: *What does a great rehab experience at this hospital look like?* Write it down on a sheet for the research team to collect with your drawings to pin to the walls.

In the same project, the design team used the medium of drawing and visualisation to innovatively imagine and share current practice and ideas for change. The team asked participants to *create – draw – a comic strip* to illustrate a typical rehabilitation patient journey between two facilities: before, during, and after admission to hospital,

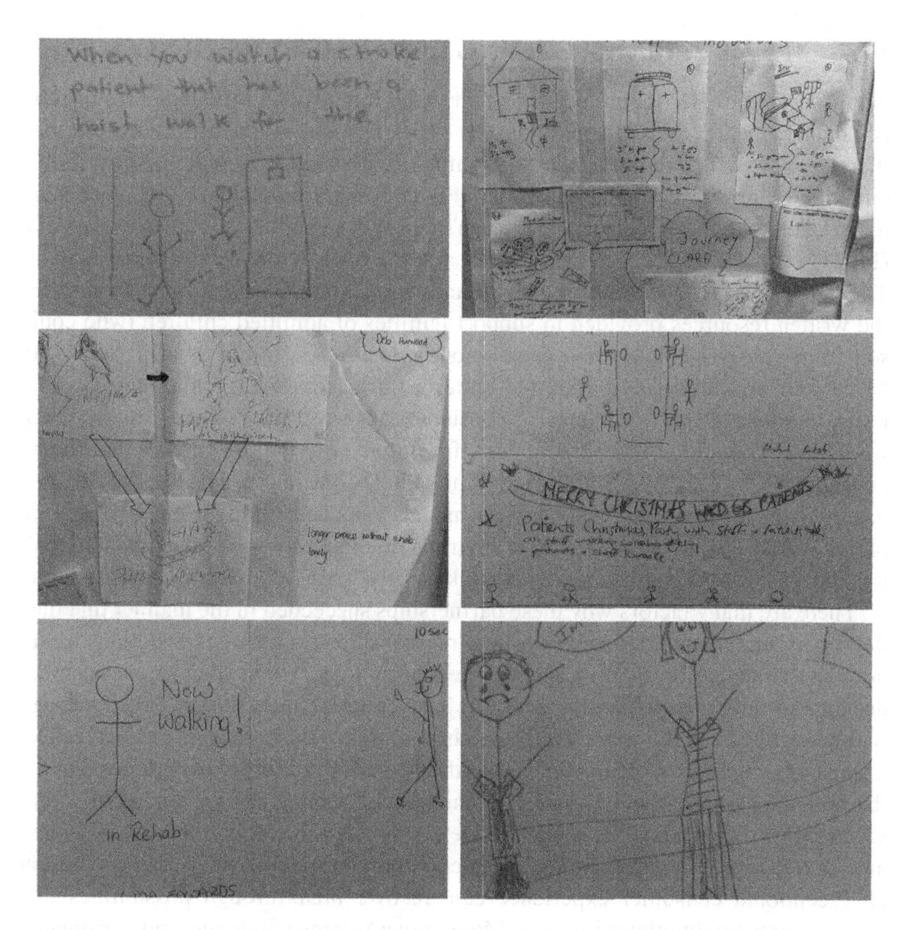

FIGURE 17.3 Drawing 'exceptional moments' and journey mapping in comics.

using the provided personas. This was, in essence, journey mapping through comics. In a playful and engaging manner, the comic strips illustrated key touchpoints in the patient journey – and once the team pinned the comics to the walls, all participants were encouraged to comment on/add scenes to other groups scenarios by pinning up 'callout' cards (writing or drawing). Whether it is sketching a moment of exceptional practice or creating a comic-inspired journey map, the use of drawing enhances creativity, collaboration, and communication, while also creating a more engaging and enjoyable design process – especially in cross-cultural contexts.

In the context of design partnerships in healthcare, designers seemed to naturally occupy and move between these dual roles of creative professional and the facilitators of the innovation process, and the best outcomes seemed to emerge from teams created sufficient space for designers to be able to work in that manner – often using creative visuals to enhance creative innovation.

17.7 THEME 2: TRANSFORMING CONSUMER EXPERIENCES

A key success of HEAL projects was that many initiatives, in the language of project management, 'overdelivered' on the brief. Healthcare teams often described their shock at reaching out to the programme for assistance around one priority but quickly finding support for a substantially larger programme of work. A piece work undertaken with the Queensland Children's Hospital's Paediatric Intensive Care Unit (PICU) is illustrative of the experience shared by many teams who worked with us. The PICU team reached out for a small amount of graphic design input to enhance the written resources provided to support families of admitted children (who could sometimes be in hospital for over six months). However, initial consultations with PICU staff and families of admitted children surfaced several additional opportunities to enhance the experience of consumers in ways that resource development would not. The design partnership grew from that early inquiry into an end-to-end consumer experience redesign programme (PICU Innovation, HEAL, 2023d) that has led to the reconfiguration of services, models for consumer involvement in service planning, reorganisation of the PICU environment, incorporation of augmented reality, and many more innovations. This work has also garnered international attention.

There are many factors why these partnerships succeeded in the manner that they did, but a fundamental reason why these relationships *exceeded* expectations has to do with that nature of the mindset designers bring to questions of consumer-centred design and how this differs from a more conventional understanding of co-design within healthcare. Like most healthcare change initiatives, co-design projects reflect a spirit of pragmatism. This means that the goals for consumer experiences model the principles of iterative improvement (small achievable goals scaffolded over previous gains) rather than excellence. While this is often the reality of what is achievable in a resource-constrained environment, consideration of what might represent an exceptional consumer experience can uncover breakthrough possibilities that project teams might otherwise miss. Many project teams (clinicians and consumers) described their experience of collaborating with designers as 'enabling', 'empowering', and 'transformative'. By asking non-critical 'why not?' questions, designers were able to disentangle assumptions that were limiting innovation. Admittedly, the creative capabilities of designers to rapidly prototype new classes of solutions were often critical to translating this newfound openness into forward movement.

Of course, we need to also acknowledge that consumers and clinicians are busy, tired, and very often change-fatigued, sometimes cynical from past experiences of participating in superficial consultation where their thoughts, input, and feedback was not reflective in the final decisions made. While many HEAL design projects were led by frontline teams, several initiatives sat within organisational priority areas and the design activity would represent a phase within a larger programme of work. In these situations, design teams learnt to be careful in managing expectations – encouraging project leaders to approach design workshops as collaborative explorations rather than 'magic bullets' or as interventions to address deeper cultural issues (even if design-led facilitation can be a powerful catalyst for teams embarking on journeys of cultural change). The HEAL design team were always extremely clear about the project focus, owners, process, and outcomes. Co-design workshops bring together a diverse range of stakeholders to ideate, prototype, and test solutions – typically, the pace is intense, people have differing experiences and strong opinions, and there can be conflict, disagreement, and difficult conversations. The facilitator's role, therefore, is to create a psychologically safe environment, in which they respect and hear all voices, ideas, and perspectives, and creativity flourishes.

In setting up a co-design workshop, the HEAL team reminded participants that this process was about thinking differently, stepping outside comfort zones, and becoming comfortable with uncertainty. Alongside respectful communication and active listening, the HEAL design team worked to develop engaging and thought-provoking activities that fostered reflection, honesty, authentic engagement, and imagination. Coupled with drawing and art (for example, having participants individually sketch or collaboratively create a visual college illustrating their vision of healthcare ten years into the future), the structure of the co-design workshop/design sprint often also purposely surfaced the diverse array of hopes, fears, preconceptions, and worries participants brought with them. In the *Hopes, Myths, Fears, Taboos (HMFT) Matrix* activity, participants would write separate post-it notes for each category: for example, list all your hopes about this proposed change, now list all your fears. The taboos category is for things we know but cannot say. In some workshops, participants did a HMFT Matrix from multiple perspectives – i.e., from the point of view of staff and then from the point of view of consumers. As a facilitator, it may be enough to surface these and pin post-it notes on large butchers' paper separately for each HMFT, or collaboratively group these and discussed them among participants at a table. HMFT is a powerful activity for surfacing fears, conflict, and tensions, and the facilitators must thoughtfully manage the process and outcomes.

Finally, as a socially active profession, designers also expressed their agency by advocating for vulnerable consumers groups within projects, sometimes challenging the remit of projects and other times encouraging project teams to accept a certain amount of risk. One such example was a short animation project to challenge institutional racism in healthcare (Animating Cultural Safety, HEAL, 2023b) led by a team of First Nations clinicians. The persuasive and provocative video generated discomfort by viewers, not to offend but to catalyse real and reflective conversations about the scale of the problem, people's hidden biases, and how clinicians could bring a more informed lens to their practice.

17.8 CONCLUSIONS

Consumer-centred healthcare improvement demands a fresh approach that prioritises consumer needs and delivers real improvements to the complex system that is healthcare. This is what a design-led approach embodied by HEAL offers: a user-centred, strengths-based, and innovative approach that enables the collaboration and creativity needed to optimise healthcare improvement. Placing the voices of consumers and clinicians at the centre, the design-led approach pioneered by HEAL offers a constructive path towards true healthcare transformation.

Design-led approaches to transforming complex healthcare systems are not a quick fix or one-size-fits-all-solution; they require a willingness to engage in a process of exploration, experimentation, and adaptation. Replicating the spirit of HEAL in other contexts, and at scale, thus requires acceptance of the 'messy' nature of change-making and from knowing that by moving out of our comfort zone we can discover creative, context-specific, adaptable solutions. Partnering with professionals who bring diverse and unfamiliar skills is not something healthcare has excelled at, but as the outcomes from HEAL highlight: partnering with the designers can (and does) lead to real and sustainable improvements in healthcare.

While it would be unfair to suggest that co-design projects underway globally across healthcare today are in some way deficient if they do not include qualified designers, in many instances, the inclusion of designers can bring extra dimensions to improvement work that are just beyond the capabilities of healthcare clinicians and consumers alone. In a sector where workforce shortages, resource constraints, and costs of service delivery continue to escalate internal best efforts, it may be time to broaden the conversation to include creative professionals as equal contributors in the grand challenge of healthcare redesign.

REFERENCES

Braithwaite, J., Churruca, K., Ellis, L., Long, J., Clay-Williams, R., Damen, N., Herkes, J., Pomare, C., & Ludlow, K. (2017). *Complexity science in healthcare: A white paper.* www.mq.edu.au/__data/assets/pdf_file/0012/683895/Braithwaite-2017-Complexity-Science-in-Healthcare-A-White-Paper-1.pdf

Cheers, J. (2023a, February 17). *Stroke mapping.* Portfolio Website. www.jessicacheers.com

Cheers, J. (2023b, February 17). *VOICED project overview.* Portfolio Website. www.jessica-cheers.com

CEQ Bridge Labs Program, Queensland Health. (2023). Retrieved February 16, 2023, from https://clinicalexcellence.qld.gov.au/priority-areas/service-improvement/bridge-labs-program

Cooperrider, D. (2017). The gift of new eyes: Personal reflections after 30 years of appreciative inquiry in organizational life. *Research in Organizational Change and Development, 25,* 81–142.

Cooperrider, D. L., & Lindsey, N. (2011). Positive organizational development: Innovation-inspired change in an economy and ecology of strengths. In G. Spreitzer & K. Cameron (Eds.), *The Oxford handbook of positive organizational scholarship* (pp. 737–750). Oxford University Press.

Cooperrider, D. L., Whitney, D. K., & Stavros, J. M. (2008). *Appreciative inquiry handbook: For leaders of change.* Berrett-Koehler.

Kannampallil, T. G., Schauer, G. F., Cohen, T., & Patel, V. L. (2011). Considering complexity in healthcare systems. *Journal of Biomedical Informatics, 44*(6), 943–947. https://doi.org/10.1016/j.jbi.2011.06.006

Macrae, C., & Stewart, K. (2019). Can we import improvements from industry to healthcare? *BMJ, 364.*

Mannion, R., & Exworthy, M. (2017). (Re) Making the procrustean bed? Standardization and customization as competing logics in healthcare. *International Journal of Health Policy and Management, 6*(6), 301–304. https://doi.org/10.15171/ijhpm.2017.35

Perry, S. J., Catchpole, K., Rivera, A. J., Parker, S. H., & Gosbee, J. (2021). "Strangers in a strange land": Understanding professional challenges for human factors/ergonomics and healthcare. *Applied Ergonomics, 94,* 103040.

QUT Design Lab. (2023a, February 16). *Using images to reduce urine contamination in the ED – the power of visual health communication – HEAL. Healthcare excellence accelerator web page and project catalogue.* https://research.qut.edu.au/heal/

QUT Design Lab. (2023b, February 17). *Animating cultural safety – HEAL. Healthcare excellence accelerator web page and project catalogue*. www.qut.edu.au

QUT Design Lab. (2023c, February 17). *Enhancing access to "just" healthcare for CALD clients: Exploring barriers to interpreter service uptake in Metro South Health – HEAL healthcare excellence accelerator web page and project catalogue*. www.qut.edu.au

QUT Design Lab. (2023d, February 17). *PICU innovation: Healthcare excellence accelerator web page and project catalogue*. https://research.qut.edu.au/heal/projects/co-desiging-a-healing-environment-for-picu-families-and-staff/

Whitney, D., Trosten-Bloom, A., & Cooperrider, D. (2010). *The power of appreciative inquiry: A practical guide to positive change*. Berrett-Koehler.

18 Good Work in the Emergency Department

A Journey through Discovery, Design, and Realisation

Elizabeth Austin

When both your parents work in healthcare, you tend to grow up hearing stories about what it is like: what the different hospitals and wards were like as workplaces, interactions with patients, diverse types of symptom clusters, diagnostic assessments, interesting diagnoses, treatments, and the challenging decisions patients and their families must make. Like many, I have had my fair share of healthcare experiences. I have been to Emergency Departments (ED) for undiagnosed acute symptoms, clinics, and subsequent annual specialist appointments. However, recent conversations have given me a deeper appreciation of both sides of the experience, the provider, and the consumer experience, and how these insights lead to a better understanding and navigation of health and the healthcare system.

For example, needing to attend an ED is typically unplanned and stressful. You or your family member is unwell and needs urgent care. The ED is busy. It is a clinical space filled with unfamiliar people using unfamiliar technology (e.g., diagnostic testing equipment, disposable packages filled with odd-shaped objects on trolleys pushed against a wall) and they perform unfamiliar tasks to an unknown timeline. It is loud with its many alarms and conversations.

> It feels overwhelming, it feels terrifying, and very isolating. You're in a bed with very thin curtain partitions around you. But you can still hear absolutely everything going on around you and can feel that chaos.
>
> **– Jack McCormack, 2022**

And it smells like disinfectant – mostly. I suspect that is unlikely that most patients feel like me: comfortable with this space, the sounds, the smells, the processes, and the language. For most, the ED experience is overwhelming: full of uncertainty about what things mean – the alarms, being asked to wait in a particular space, and the waiting time (e.g., if you have been waiting for a long time, does it mean unwelcome news for you or just a logistical delay?). Also, there is uncertainty about what will happen and what the outcome will be.

 DOI: 10.1201/9781032711195-21

Listening to my parents' experiences and those of their colleagues, working in ED can be a similar experience. The ED is busy with the sometimes-constant arrival of patients to the department, the turnover of bed spaces, and ensuring that devices are clean, functioning, and ready for use. Providing care and communicating with patients can be satisfying and energising. But it can also be stressful and frustrating. For example, ED staff cannot predict what illness or injury patients will present with, how sick they will be, how many patients, and at what time, what technology will be needed and whether it will be available, and how the composition of staff will function during a shift. There are many sources of stress and frustration for both sides of the care experience (the carer and the patient or their family).

The stress and frustration experienced by both sides hint at a core tension in emergency care between time and safety. In emergencies, decisions need to be made quickly. For efficiency, it is necessary, then, to homogenise diagnostics and care planning based on stereotypes (e.g., gender, age, cultural and language background, and physical and psychological capacity). However, unintentional cognitive (unconscious) biases that derive from stereotypes challenge patient safety because they impact patient–provider interactions, interprofessional interactions, diagnostics, and treatment decisions (Gopal et al., 2021).

> When I first started in the emergency department, I noticed that patients waiting for acute surgery were often nervous and even anxious. They wanted to know how long it would be until their surgery and this was impossible to tell them due to the waiting list in the department and the need to prioritise urgent cases.
>
> **Antonsen, a nurse in the Emergency Department at Odense University Hospital, Denmark (Taylor, 2022)**

For example, providers may change how and what they communicate (i.e., the level of medical detail) based on their cognitive biases. Explicit and implicit attitudes towards patients based on their perceived identity and capacity can perpetuate inequalities in healthcare delivery and patient outcomes. So, how can we make ED care experiences safer and less stressful for patients and their carers?

18.1 DISCOVERY

Patients experience the ED journey as a linear system. The patient ED experience narrative follows a chronological path through the ED with predefined plot points (Figure 18.1). For example, registration with the ED clerks, triage, seeing a doctor, some form of diagnostic assessment, treatment, seeing the doctor again for review, and then either being discharged home or transferred to another ward or clinical space. Among these plot points is the waiting time, which can seem unending due to the unpredictable or unknowable length. Similarly, ED staff describe their workflow as a linear process. For example, providers' stories often follow a linear pattern

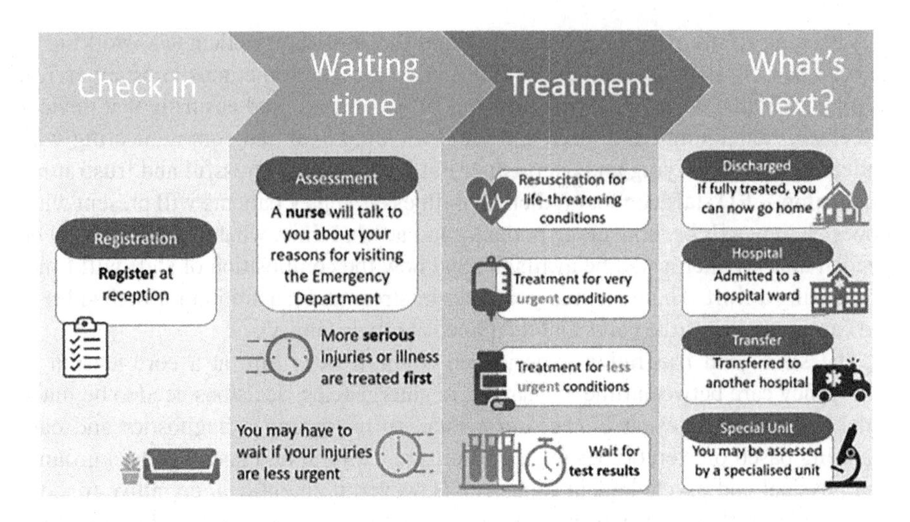

FIGURE 18.1 The plot points for the patient experience of ED.

from the patient's presentation, the doctor's assessment, what tests were ordered, what treatment was prescribed, and so on. Interestingly, the EDs performance is also predominantly measured according to this defined, chronological path (Sørup et al., 2013). For example, time to triage, time to be seen, and length of stay.

Interestingly, while the experience of patients, the way that we tell stories about the ED experience, and how ED performance is measured, all describe the patient progress through the ED in a straight line (sometimes with lots of waiting), ED work is non-linear. ED work is non-linear because it is performed in a complex, adaptive system. Each ED staff member (e.g., doctors, nurses, allied health, or professional staff) makes decisions about what tasks to perform: *Am I assessing a patient? Am I documenting? Am I advocating or linking in with specialists?* Each staff member also interacts with other staff members. A lot of that is local interaction (i.e., within the ED), but can extend to experts and services outside the ED (e.g., patient transport, specialist teams within the hospital, or community services). What emerges from their task performance and interactions is the ED system. The ED system is a product of their work, not the sum of the parts (tasks and interactions) (Ackoff, 1994). We cannot understand the ED work system by looking at an individual task, profession, or interaction. The interactions between individuals in the performance of tasks within the physical environment mean that the ED system is dynamic and complex. The ED system, independent of individual staff and patients, is robust and adaptive. Therefore, we must use complex systems models and tools to understand the complexity of this adaptive system.

The Systems Engineering Initiative for Patient Safety (SEIPS) model provides a framework for understanding the ED work system and includes five pillars: the technology and tools (e.g., beds, chairs, computers, diagnostic and treatment devices), the tasks (e.g., administering medication, handwashing, assessing, and streaming), the physical environment (e.g., the layout, cleanliness), the people (e.g., knowledge, roles, skills), and the organisation (e.g., communication, culture, governance) (Carayon et al., 2006). These pillars interact during clinical (e.g., care) and non-clinical (e.g., admin) processes, resulting in outcomes for patients (e.g., safety, timely diagnosis, or care quality) and staff (e.g., job satisfaction, safety, health, and turnover). The performance of one task in ED can consist of many linked workstations (e.g., with the patient, the diagnostic device, the nurses' station computer, the printer, and the procedures trolley) where workflow might frequently vary, as might interactions with various staff members.

ED staff must be flexible in their decision-making to manage the complexities during their performance in the ED work system. This means that ED staff and patients contribute to the patient journey using a variety of tools and technology to perform a range of tasks in the ED physical space within the hospital organisation. ED staff make decisions about the order in which they perform work tasks and, to a certain extent, *with whom* they interact to complete the tasks. These decisions influence the patient journey and offer several outcomes. There are, of course, policies, guidelines, and protocols that describe or prescribe how work should be performed in ED. These documents describe work-as-imagined (WAI) (Braithwaite et al., 2016) and can be quite overarching in their prescription of work, allowing each ED to adopt the work prescription in a way that will fit their local environment and resources. But these documents also describe ED work as linear and do not necessarily account for the non-ideal and non-linear state in which the system might exist at any one time.

Work as it happens is called work-as-done (WAD) (Braithwaite et al., 2016; Shorrock, 2017). In ED, WAD does not take place in an ideal system state. Unexpected things happen, tasks take longer than expected, or the situation changes. ED WAD is living everyday work, and because unexpected things happen, it can differ from WAI considerably (Clay-Williams et al., 2020). For example, devices not working, trolleys being out of stock or not being left where they should be. When the context or situation differs from WAI, ED staff must adapt their activities to achieve their care delivery goals. These adaptations are often referred to as 'workarounds'. In the ED, it might be that staff look for a fully stocked trolley, a functioning device, or someone with the correct qualifications to perform a task.

In complex adaptive systems, there must be more than one way to perform a task (i.e., performance variability, adaptive capacity, or workarounds) (Naikar & Elix, 2016). Without the capacity for performance variability, systems become fragile. Fragile systems are sensitive to expected and unexpected events, resulting in system failures. Sometimes the workarounds reveal a better way of working, leading to a revision of procedures (i.e., WAI). Sometimes the workarounds reveal gaps in the system that may be a result of a unique series of events, or unexpected consequences of system changes, resulting in something gone wrong (or deviating from an intended or expected outcome) (Wachs et al., 2016). Understanding WAI and WAD, and ED staff's adjustments are critical to establishing a shared understanding of why things usually go right and, occasionally, why they go wrong.

There are complex systems tools that allow us to capture the non-linear nature of ED WAD and performance variability to accurately describe the ED system. For example, Functional Resonance Analysis Method (FRAM) is a method to analyse work activities and understand the drivers behind performance variation (Hollnagel, 2012). We can use FRAM to produce a model or visual representation that describes

WAI or WAD in terms of the functions involved (i.e., the things necessary to achieve an aim or perform an activity). In this way, FRAM provides a means for developing a shared understanding of how the ED system should work (i.e., WAI), and how it currently works (i.e., WAD) (Clay-Williams et al., 2015). Another method for understanding the non-linear nature of ED work is the Cognitive Work Analysis Framework (CWA) (Austin et al., 2021). CWA is a suite of five models that allow us to explore complex systems like the ED work system, including an overarching system description (the Work Domain Analysis) and models that can be used to describe *what*, *where*, *how*, *who*, and the different cognitive capacities required for how work is done or developing ideas for how work could be done (Austin et al., 2022). Crucially, both models allow us to explore and make explicit the *why* – what is the reason, cause, or purpose for the way work is performed.

FRAM and CWA allow us to capture the complexity and nature of work; however, there are some challenges in applying these tools to a complex, dynamic system like ED. The ED system is in a constant state of flux. The resources available, the combination of staff members (e.g., skill, knowledge, experience, attitudes, or burnout), and the patients (e.g., illness, injury, complexity, acuity, health literacy, language, and culture) in the department can change significantly throughout a single shift, let alone over the time it takes to create a model. So, the models need to be living (i.e., updated regularly) to ensure that they remain relevant. The models that we create (e.g., FRAM, CWA) are complex and require us to collect data in the ED. This includes extracting information from policy and guideline documents, spending time in the ED to observe how work is performed, interviewing staff, or running structured workshops in which models are developed (Austin et al., 2021). But data collection in the ED is challenging. EDs are time-critical care delivery environments where staff do not always have the time to contribute to the development of complex models. Despite these challenges, the development and use of tools like FRAM and CWA are critical to establishing a shared understanding of the non-linear characteristics of the ED work system, and the gaps between WAI and WAD.

If we are to improve the ED care experience for patients and those who care for them, then we need tools that accurately reflect the complex, dynamic, non-linear characteristics of the ED system. Complex systems tools such as FRAM and CWA allow us to describe and understand complex, dynamic, non-linear systems. Once we have a better understanding of the ED system, we can design innovations to improve the ED care experience.

18.2 DESIGN

Before we can talk about designing innovations in care, we need to understand the *why*, *who*, and *what*. Why design innovations? The ED system can cope with large variations in demand and variable resources. The constant adaptions and workarounds accommodated by staff mean that patients receive safe, timely, and quality care. So why might we want to redesign or design ED systems? And *who*? Who designs innovations? It would make sense that ED staff are the best equipped with knowledge about their needs and preferences based on their experiences. But ED staff do not have time to devote to the development of innovations. So, who needs to be involved in the redesign or design of ED systems? And why, after all these years of quality improvement projects, have we not seen an improvement in our system? Finally, what innovations are we designing? Are we going to continue to refine parts of the system or are we at a point where we can consider a shift in thinking about care delivery from a disease focus to a wellness focus?

Why design innovations? One reason is staff frustration. There is a trade-off among safety, time, and quality in the performance of any task in the ED. For example, clinical (e.g., comprehensive examination, or taking blood) and non-clinical tasks (e.g., comprehensive documentation, or data entry) take time, but this means there is less time for other critical tasks or patients.

Unable to offload patients from ambulances, emergency staff don't feel they are able to meet the needs of their patients with emergency medical officers reporting in an inquiry into ambulance ramping. 'We work in an environment that we don't see as conducive to good medical care', Dr Tadros (Shams, 2022).

The tensions among safety, time, and quality goals create frustrations for staff. So, staff might look for innovations to reduce those tensions and improve the quality of care that they can provide. Another reason might be that the ED system is perceived as not performing (i.e., according to key performance indicators) and there might be pressure from stakeholders to improve care delivery (internal stakeholders such as management, or external stakeholders such as the community). Or a patient might experience harm in the ED, driving staff, and hospital stakeholders to address the factors that might have contributed to that experience. The design or redesign of ED systems looks to improve care delivery for staff and patients, but there can be different triggers or drivers for that innovation.

Who does the (re)designing? Often the (re)design is done by frontline ED staff. The (re)designs range in scale from small ones, such as changing the type of disposable product used to a recyclable version, introducing artwork (e.g., murals) in the environment, or reorganising a clinical space to optimise task performance (i.e., ergonomic) through to larger scale innovations, such as the development of a new model of ED care (e.g., new patient pathways, or new protocols for care). Tensions remain for ED staff, however, in the division of time between caring for patients and developing and implementing their (re)design innovations.

The Clinical Services Redesign programme in NSW uses a 'whole-of-system' approach to address things in the patient journey that might get blocked, duplicated, or complicated. For example, 'putting blood tests from ED into different coloured bags to the rest of the hospital so pathology know which ones to fast-track' . . . 'its genuine grass roots approach' . . . 'the changes were made not by bureaucrats, but nurses, doctors, physios, trolley attendants and consumers' (Jakubowski, 2005).

ED staff are further constrained by aspects of the system that cannot be changed (e.g., number of beds or computers), their exposure to other EDs (i.e., knowledge about how other EDs do things), and their knowledge about how to go about it or agency (i.e., how to manage change or the authority to do so). So, while staff might be encouraged to innovate, they often do not have formal training in how to do it, or the time, authority, or resources to develop and implement effective change.

Quality improvement projects are intended to improve care delivery for staff and patients. And quality improvement projects are often designed by those with the most knowledge and experience about how care is delivered in the ED. But these improvement projects are doomed to failure. They are doomed to fail because they typically address a part of the ED system (e.g., the suture trolley, the organisation of the storeroom, or the procedure for dealing with broken bones). Yes, the functioning of the parts within the ED system impacts the functioning of other parts within the ED system. The parts are interconnected and interdependent. For example, developing and implementing new procedures for specific types of presentation (e.g., broken bones, back pain, viral symptoms) will influence the performance of tasks that share resources with that new procedure. For instance, staff workflow, the availability of consultation rooms, and how patients experience care in the ED. In this way, changing one part of the system changes other parts of the system. But it will not improve the performance of the entire system. ED system performance depends on the fit: how the parts within the ED system act together. So, instead of looking to improve parts of the system, we should be talking about the entire system and how the parts interact.

It helps to use tools like FRAM (Hollnagel, 2012) and CWA (Austin et al., 2021) when looking to design or improve the entire system, particularly systems that are complex and in a constant state of flux. For example, FRAM models describe what happens in terms of functions. The FRAM visual description of how a set of activities is carried out helps us to understand the possibilities for performance variation as well as the interactions with other system elements. As such, FRAM offers a way to understand outcomes that are emergent and non-linear, as well as the drivers behind performance variation. FRAM can also provide insight into what could be done to address the barriers to task performance. CWA is a more powerful design tool for complex systems. The phases of the framework support the development of innovations from the big picture and address *why* are we doing this. This extends to the granular *who*, *what*, *where*, *how*, and with *what resources*. Both FRAM and CWA models can be used as living models – updated as the system evolves. A system that is in constant flux presents evolving challenges between the intervention and the context as the innovation and implementation strategy must account for the level of constant change in the system. Any system innovation needs to be underpinned by an assumption of continuous change for its long-term success.

So why aren't all EDs using tools like FRAM and CWA to design and implement innovations? While FRAM and CWA are powerful system tools, that power comes at a cost. It takes time to learn how to develop and use these models. Time which ED staff might not have. EDs would need to then collaborate with the developers of FRAM and CWA models. These human factors experts would need to collect data to inform the models. The models, once developed, need to be maintained to be kept 'living', because the ED system is constantly changing. The models can be used across different EDs; however, they need to be validated in the local context. The models are complex, and while reflective of the ED's complex adaptive system, the complexity might make them difficult to use or deter ED stakeholders from applying them. As a result, ED staff are not the only people involved in the (re)design and innovation of our ED systems.

Ideally, the design process of innovations to the ED system, such as new processes, protocols, and models of care, includes provider and consumer consultation facilitated by a systems designer (e.g., human factors systems design researcher or consultant). Including providers and consumers in the design of ED systems is critical for the design of effective and sustainable workflow. Providers have distinct roles and goals within the ED system resulting in unique needs, preferences, behaviours, and experiences. Similarly, consumers' needs, preferences, behaviours, experiences, and outcomes are not homogeneous. The role of the systems design researcher (either external to the organisation or embedded within the organisation) is to lend scientific rigour to the work with providers and consumers to understand their problems and collaboratively develop ways to improve the system. Incorporating diverse perspectives into the design process facilitated by a systems designer or researcher generates clinically effective and sustainable innovations that are aligned with system, staff, and consumer needs.

However, actively partnering with hospital stakeholders (i.e., ED doctors, nurses, allied health, professional staff, and ED leaders) and consumers to collaboratively develop innovations is challenging. Facilitated co-design requires workshops, focus groups, or interviews are used to develop and refine ideas. Time and expertise (i.e., research, tools, and change management) constraints are significant barriers to provider and consumer consultations. While establishing and leveraging collaborations with experts in research, applying scientific tools and methods can reduce barriers to co-design and help realise innovations.

> It wasn't until the emergency department team – from the clerical staff to the head of department – got together to map how patients experience the system, that we realised how confusing it was.
>
> **– Ben Tovim (Bisset, 2008)**

What innovations are we designing? Care delivery and the innovations we implement typically focus on disease (i.e., pathogenic) – diagnosing and treating based on disease. For example, there are protocols and pathways for cardiac patients who present to the ED that improve the likelihood of good health outcomes. A disease focus makes sense for ED as it's where you go when you need urgent critical care. However, patients are more than their disease or illness. They also have different health goals and resources. There is a need, therefore, for a shift in ED system design from a disease focus to a holistic approach. For example, the Salutogenic Framework emphasises the importance of focusing on peoples' resources and capacity for health behaviours (Lindström & Eriksson, 2005; Mittelmark et al., 2022). A health system designed from a salutogenic perspective would centre care delivery around the goals of the patient as well as their ability and capacity to manage. An ED system that is designed from a salutogenic perspective would not only treat an individual's acute needs but also support patients to manage and respond, connecting them with community-based resources. In addition, the care would consider the individual's health, wellness, cultural, linguistic, family, health literacy, literacy, occupational, social, and economic context. In this way, ED care could be delivered to patients in a way that is safer for patients and meets their holistic health and wellness needs.

Intentional ED system design is critical for patient and staff safety. The development of innovations in the ED system that align with stakeholder needs requires the representation of staff from different professions and diverse consumers, as well as complex systems experts to be part of the design conversation. Complex systems tools like FRAM and CWA can help support effective sustainable system design but can require a large investment of time and skill development. Effective and sustainable ED system designs align with system, staff, and patient needs.

If we are to improve the ED system, manage worker frustrations or cognitive loads, patient care outcomes, and subsequently the care experience for patients and those who care for them, then we need to identify measures (i.e., clinical, system, and experience outcomes) that accurately reflect the impact of the innovation and implement it. Complex systems tools such as CWA can support performance measure selection and implementation in complex, dynamic, non-linear systems. Once we have designed an innovation to improve the ED care experience, we can realise that innovation by implementing, measuring, and refining it.

18.3 REALISATION

We design and implement innovations in care delivery to improve care quality, outcomes, patient and staff experiences, and system performance. But how do we know if our intentions have been realised? It makes sense to use existing measures to decide if our innovation has been successful or not, but what do we do when the measures do not tell us what we need to know? The ever-evolving work environment (e.g., workforce composition changes, pandemics, modern technology, or new models of care) means that designing and implementing innovations in the ED is a continuous journey. How do you make a change sustainable or an adaptive intervention that evolves with the system as the system changes? And when is a suitable time to measure if the intervention has been successful? And what happens after? How do we go from one intervention in one context to the standardisation of care delivery and the formal establishment of the intervention as a recognised procedure with rules (i.e., policy)?

There are already a lot of measures for ED system performance (Austin et al., 2020). For example, time to triage, time to treatment, and the number of presentations. So, we could use these existing measures to decide if the innovation of our system has been successful. However, these measures are typically volume based. They measure the level of activity in an ED and how efficient it is. But these measures do not tell us about value, or about how effective our ED system is, or how effective our innovation has been through qualitative measures. *Efficiency* tells us how well something is done (per productive measures); *effectiveness* is about how useful it is. Systems thinking and system models like CWA can help us identify existing measures to assess our innovations per efficiency measures and develop new measures that allow us to evaluate whether our service intention is being achieved (i.e., effectiveness). For example, measures for quality, patient-centred care, and teamwork.

Systems change over time. The ED system is in a constant state of flux. The workforce composition changes and contemporary technology are being introduced into care processes, along with new models of care. System changes impact the sustainment of innovations. When we co-design systems with our stakeholders, we need to ensure that our innovations and the work environment align and that we account for changes to occur over time (Chambers et al., 2013). Effective sustainable innovations require a whole-of-system approach, rather than addressing system parts. For example, addressing how a medication is dispensed in the ED, independent of the broader system, will not improve the performance of the ED system and may not necessarily reduce medication errors because we will have failed to account for how medication dispensing interacts with other work tasks and roles (i.e., the interaction with other aspects of the ED system).

While systems change over time, we need to design for now rather than the future and we need to design for what we want, rather than what we do not want. This sounds basic, but it is not something we typically do well. Designing for what we want means you are designing for what you want to do or the outcome that you want (e.g., good community health and well-being, and quality care outcomes), whereas designing for what we do not want is designing for what you do not want to do or outcomes you do not want (e.g., medication errors, adverse events). The key here is understanding that designing for something that we do not want (e.g., adverse events) will not necessarily get us what we do want (e.g., quality care outcomes). This exemplifies the ideas around innovative, design thinking (a holistic view) versus risk management (which can be reductionist in view).

Given the constant state of flux within the system and the broader community context within which the ED sits, it is likely that determining if the innovation is the cause of the improvement (or deterioration) in system performance is going to be difficult. We should be measuring how our ED system is performing all the time (before, during, and after), and examining the internal and external events that are likely to impact the system (the context of performance). This will help explain why the system performs the way that it does. We need to pair the timing of measurement with what we are measuring and transition from only measuring efficiency to measuring both efficiency and effectiveness. This will enable us to evaluate whether the purpose of the system is being pursued.

Standardising care delivery through establishing procedural rules and regulations (for how to provide care – policies and standards) improve patient safety. Standardising care through policies and procedures reduces unnecessary variation in care delivery. Variations in care delivery and clinical decision-making stem from differences in medical training, experience and knowledge, new evidence (research), patient characteristics, clinical environment constraints, and the accessibility of other professional expertise, to name but a few. For example, an ED registrar trained in Australia with an interest in toxicology who recently attended a training session on anaesthetics may be able to identify delirium as a post-operative symptom, whereas an ED registrar trained in Australia with an interest in cardiology may identify delirium as an expression of a mental health crisis. As a result, governance bodies look to establish models of care, clinical pathways, and procedures as tools for standardising care. Governance bodies work with researchers and clinicians, and more recently with patients to establish processes at their highest level, focusing on decision points to guide care.

Like with the design, the implementation and sustainment of innovations need to be led and owned by the ED. The introduction of change without the ED team leading or owning the change will irritate the system, leading to the ED system rejecting the change. For the scale and spread of innovations to other EDs, resources are needed, including expertise and tools such as the CWA and FRAM to support their implementation and maintenance. We must use measures of efficiency and effectiveness along the way because our goals relate more to the quality of care and the volume of care delivered. The standardisation of care should reduce the degree to which each ED is solving the same problem by collectively sharing solutions, and it should improve patient outcomes by minimising the impact of variation on care delivery.

18.4 WHERE DO WE GO FROM HERE?

To make care safer for patients and less stressful for those that care for them, we need two things. First, we need many people to be part of the conversation. This includes researchers, systems experts, staff (clinical and professional staff at various levels), and patients. The point of these conversations is to support decisions about system design by providing information about what the evidence base (research) is, what the preferences or values are of the stakeholders, what the contextual constraints or circumstances are, and what the stakeholder's experiences and judgements are (Briner et al., 2009). The second thing we need is complex systems tools. Complex systems tools, such as the CWA and FRAM, provide a platform for conversations: tools to support the development of a shared understanding of the system, and tools to support the design and implementation of changes that account for the complexity of the system.

IMPLICATIONS FOR PRACTICE

- Consultation – Many people to be part of the system design conversation – providers, including professional staff, technology developers, manufacturers, researchers, systems experts, and patients. A diverse group of patients must be part of the system design conversation. It is only then that we will be able to ensure the system is meeting the needs of the community it is serving.
- Embedded learning role – A role that supports the application of systems tools and works collaboratively with staff, patients, and researchers to develop quality improvement initiatives
- Embedding systems tools to support systems thinking: Tools such as the CWA to support the development of a shared understanding of the system

and that support the design and implementation of changes that account for the complexity of the system.
- Training and ongoing training in the use of systems tools for their introduction and ongoing use in ED system design and management.
- A shift in thinking and culture, from a focus on parts of the system to interactions among parts to improve care delivery, and from designing for what we don't want to what we want.

REFERENCES

Ackoff, R. L. (1994). Systems thinking and thinking systems. *System Dynamics Review*, *10*(2–3), 175–188.

Austin, E. E,, Blakely, B., Salmon, P., Braithwaite, J., & Clay-Williams, R. (2021). Identifying constraints on everyday clinical practice: Applying work domain analysis to emergency department care. *Human Factors*, *64*(1), 74–98. https://doi.org/10.1177/0018720821995668

Austin, E. E., Blakely, B., Salmon, P., Braithwaite, J., & Clay-Williams, R. (2022). Technology in the emergency department: Using cognitive work analysis to model and design sustainable systems. *Safety Science*, *147*, 105613. https://doi.org/10.1016/j.ssci.2021.105613

Austin, E. E., Blakely, B., Tufanaru, C., Selwood, A., Braithwaite, J., & Clay-Williams, R. (2020). Strategies to measure and improve emergency department performance: A scoping review. *Scandinavian Journal of Trauma, Resuscitation and Emergency Medicine*, *28*(1), 55. https://doi.org/10.1186/s13049-020-00749-2

Bisset, K. (2008). Blood works better in colour. *Weekend Australian*.

Braithwaite, J., Wears, R. L., & Hollnagel, E. (2016). *Resilient health care, vol. 3: Reconciling work-as-imagined and work-as-done*. CRC Press. https://books.google.com.au/books?id=MDQNDgAAQBAJ

Briner, R. B., Denyer, D., & Rousseau, D. M. (2009). Evidence-based management: Concept cleanup time? *Academy of Management Perspectives*, *23*(4), 19–32. https://doi.org/10.5465/amp.23.4.19

Carayon, P., Schoofs Hundt, A., Karsh, B. T., Gurses, A. P., Alvarado, C. J., Smith, M., & Flatley Brennan, P. (2006). Work system design for patient safety: The SEIPS model. *Quality and Safety in Health Care*, *15*(Suppl 1), i50. https://doi.org/10.1136/qshc.2005.015842

Chambers, D. A., Glasgow, R. E., & Stange, K. C. (2013). The dynamic sustainability framework: Addressing the paradox of sustainment amid ongoing change. *Implementation Science*, *8*(1), 117. https://doi.org/10.1186/1748-5908-8-117

Clay-Williams, R., Austin, E., Braithwaite, J., & Hollnagel, E. (2020). Qualitative assessment to improve everyday activities: Work-as-imagined and work-as-done. In *Transforming healthcare with qualitative research* (pp. 71–82). Routledge.

Clay-Williams, R., Hounsgaard, J., & Hollnagel, E. (2015). Where the rubber meets the road: Using FRAM to align work-as-imagined with work-as-done when implementing clinical guidelines. *Implementation Science*, *10*(1), 1–8.

Gopal, D. P., Chetty, U., O'Donnell, P., Gajria, C., & Blackadder-Weinstein, J. (2021). Implicit bias in healthcare: Clinical practice, research and decision making. *Future Healthcare Journal*, *8*(1), 40–48. https://doi.org/10.7861/fhj.2020-0233

Hollnagel, E. (2012). *FRAM: The functional resonance analysis method: Modelling complex socio-technical systems*. Ashgate.

Jakubowski, L. (2005). Hospital gridlock. *Weekend Australian*.

Lindström, B., & Eriksson, M. (2005). Salutogenesis. *Journal of Epidemiology and Community Health*, *59*(6), 440. https://doi.org/10.1136/jech.2005.034777

McCormack, A. (2022). Youth suicide report urges reform to emergency department care for young people in distress. *ABC News*. www.abc.net.au/news/2022-08-12/suicide-prevention-australia-report-recommends-ed-overhaul/101323008

Mittelmark, M., Bauer, G. F., Vaandrager, L., Pelikan, J. M., Sagy, S., Eriksson, M., Lindström, B., Meier Magistretti, C. (2022). *The handbook of salutogenesis*. Springer Nature. https://doi.org/10.1007/978-3-030-79515-3

Naikar, N., & Elix, B. (2016). Integrated system design: Promoting the capacity of sociotechnical systems for adaptation through extensions of cognitive work analysis [hypothesis and theory]. *Frontiers in Psychology*, *7*. https://doi.org/10.3389/fpsyg.2016.00962

Shams, H. (2022). *NSW inquiry into ambulance ramping told patients "dying unnecessarily"*. www.abc.net.au/news/2022-10-05/nsw-patients-dying-unnecessarily-ambulance-inquiry-told/101504524

Shorrock, S. (2017). *The varieties of human work*. www.safetydifferently.com/the-varieties-of-human-work/

Sørup, C. M., Jacobsen, P., & Forberg, J. L. (2013). Evaluation of emergency department performance – a systematic review on recommended performance and quality-in-care measures. *Scandinavian Journal of Trauma, Resuscitation and Emergency Medicine*, *21*(1), 62. https://doi.org/10.1186/1757-7241-21-62

Taylor, T. (2022). *Music in a pillow can calm pre-op patients*. www.abc.net.au/radionational/programs/healthreport/music-pillow/101570426

Wachs, P., Saurin, T. A., Righi, A. W., & Wears, R. L. (2016). Resilience skills as emergent phenomena: A study of emergency departments in Brazil and the United States. *Applied Ergonomics*, *56*, 227–237. https://doi.org/10.1016/j.apergo.2016.02.012

19 It Happened on a Wednesday

A Reflection on an Emergency Department Admission

Sara Pazell

On a Wednesday, a day that should have been innocuous like most mid-weekdays unfold, I woke and felt numb in the face. It was a jolt to my sensibilities when I looked in the mirror: the right side of my face had dropped. Not just a little, but a lot. The entire right side of my face seemed dead: frozen in time. The forehead and brow were sunken and would not lift, despite my conscious efforts or spontaneous attempts at a smile. The right nostril caved in. The right cheek was sunken, and the lips were downturned. This was not my face. The experience was beyond one incurred during natural ageing: seeing a reflection marred by more wrinkles at the edges of the eyes or the greyed roots at the temples that defy the marvel of a curious and optimistic mind teeming of youth beneath the façade of flesh. This experience followed new-onset prodrome migraine experiences (new in the last two to three months, but this time the symptoms deviated). I had woken with a right-sided Bell's palsy on a sunny Wednesday morning.

I managed to see a general practitioner in the afternoon who referred me to the emergency department. I attended the hospital late that afternoon. What follows is my response to their post-discharge patient survey:

> My experience was this: A quick admission because of stroke concerns. Had I not been a healthcare professional, the process would have been more alarming because the calls for 'stroke specialist' were audible. Consider using a colour code term to replace the calls down the hallway for a stroke specialist while the patient lies in wait, creating new narratives in their mind based on the frenzied, panicked state of the staff.
>
> Luckily, I brought a water bottle and warmed soup before arriving and I helped myself during the waits. In my 5 hours or so there, I was not once offered a drink of water. I used a jacket for a pillow most of the time because, for hours, a pillow could not be found by the staff.
>
> The observations and initial gross medical review were negative for stroke, but I was referred for a computerized tomography (CT) scan to investigate brain haemorrhages. I explained my need to provide care and support for my son (owing to sole parent obligations) and repeatedly asked for an estimated time for the CT scan procedure.

DOI: 10.1201/9781032711195-22

I asked again at the conclusion of the staff shift change over. I understand triage needs for trauma cases, but I needed to plan for my son's care. The nursing staff could not advise me of an expected time. After several hours of waiting, I asked to self-discharge. This caused a sudden review of my case, and it was discovered that the neurological registrar, the doctor who saw me at my admission, failed to order the CT scan, so I had been waiting in vain. This delay costed a hospital bed, my time, and caused me concern for my son. The (new) doctor on shift promptly expedited the CT scan request and I was seen 10–15 min later. There were more delays until the doctor was available to provide discharge scripts, and I contributed to the care planning because of research I was undertaking. Had I not been a confident self-advocate, this could have turned into an overnight stay. I also asked to walk versus wheel around passively, pushed in the bed by an orderly, but I was told that this was not the process (and I would 'lose the valuable bed'), so the angst of being in a sick role was perpetuated in this system.

There was a woman with mental health disorders howling, moaning, and sometimes screaming just across the hall the entire time. There were no private or protected areas with acoustics management for this person, the staff, or other patients.

The structural pillars in the middle of the emergency department (ED) hallways caused challenges for orderly staff to manoeuvre patients on beds and supply carts. I am a human factors researcher and practitioner, and a human factors systems design review is sorely needed of this ED to help overtaxed staff and patients enjoy a better, more efficient experience. Please call me if (Organisation Y) is prepared to make these improvements. I believe that (Organisation Y) just disbanded it's research collaboration on design improvements led by a human factors researcher and, if this is true, this is an utter shame. Review of EDs with human factors approaches is essential, if my experience represents that of others (and I hear that this is the case). The staff are trying their best in an utterly poor system.

This is a patient perspective. It is a true story, with only minor edits made to the hospital survey response for the purposes of the editorial requirements of this book. What was not shared in the letter was the query that I made of the initial 'stroke specialist' neurology team registrar on call. Migraine headaches are complicated prodrome and postdrome symptoms and, in my case, this included visual disturbances, excessive thirst, dizziness, nausea, and sensorimotor changes to the face. I asked about the correlation or causation of migraine and Bell's palsy, the case I presented with causing face drop on one side. I was told 'no association whatsoever', by the intake neurology team registrar, later contested by the second doctor. Yet, neither had time to do a proper review of my history or listen to my experiences leading up to my clinical presentation that evening. In fact, the literature suggests that migraine increases the risk of Bell's palsy and, in some cases, it may double the risk (Kim et al., 2019).

This story exemplifies the need for work redesign in EDs; thus, the importance of Elizabeth Austin's research on human factors approaches to improve the performance of EDs.

REFERENCE

Kim, S. Y., Lee, C. H., Lim, J. S., Kong, I. G., Sim, S., & Choi, H. G. (2019). Increased risk of Bell palsy in patient with migraine: A longitudinal follow-up study. *Medicine*, *98*(21), e15764. https://doi.org/10.1097/MD.0000000000015764

20 Reflections on Work Design

Kym Bancroft

As someone who has worked as a national head of work health and safety in organisations and a health and safety regulator across various industries, I have had the opportunity to witness and practice different methodologies in the pursuit of sustainable workplace outcomes. Just like the authors of these chapters, I have a strong appreciation of the value of a human-centred approach and have faced the challenges of integrating design capacity into daily work practices. That is why the ideas and concepts presented in these chapters are visionary, bold, and incredibly inspiring for health and safety professionals (like you and me).

One of the standout proposals comes from Pazell, who takes a proven methodology and suggests formalising this approach through the role of Chief Work Design Strategist (CWDS). This idea catapults us into the future, going beyond traditional health and safety roles to achieve resilient and sustainable futures through strategic work design. The thought of embracing such a role is enticing, cutting-edge, and empowering.

During my experience, I've come across roles that hint at similar ideas, but the CWDS role takes it a step further by combining the best practices from various theories and methodologies. It recognises that all areas of a business can benefit from good work design. I remember when I implemented a contemporary event using learning methodology for safety-related events. Soon enough, the finance team, human resources team, and representatives from the call centre knocked on my door, wanting to implement the same learning approach for their challenges. Unfortunately, due to resource constraints, we could not assist them in implementing the methodology across the organisation. However, a CWDS would have been a game changer in such situations because they could have spearheaded the implementation organisation-wide.

But what would it take for organisations to create and appoint a role like the CWDS? Firstly, the organisation needs to understand and acknowledge the importance of design in improving human performance and business outcomes. This understanding can be fostered by presenting compelling cases for change, such as sharing real-life examples or case studies from industry. Secondly, it's crucial for the organisation to clearly align the specific responsibilities and objectives of the CWDS role with its overall strategy and goals. Allocating adequate resources and budget to support the activities of the CWDS, including managing a design strategy budget and project portfolio, is also essential.

DOI: 10.1201/9781032711195-23

The success of the CWDS role relies heavily on recruiting individuals with deep knowledge and experience in design strategy. Additionally, having top-level support – such as the Board, Chief Executive Officer, or an executive sponsor who is enthusiastic about cultivating a culture of innovation, collaboration, and openness to change – is crucial in integrating the CWDS into the executive leadership team. This support empowers the CWDS to implement and experiment with novel approaches to work design.

However, creating and implementing the CWDS role may face barriers. There might be limited industry awareness or understanding of the value of design in work practices, which can lead to resistance and reluctance in adopting innovative approaches. This, in turn, makes it challenging to secure budget and approval for the role. To overcome these barriers, it's crucial to educate key stakeholders about the benefits of design and emphasise its importance in achieving organisational goals. Demonstrating the potential return on investment from design initiatives can also help in securing funding and resources.

One effective strategy is to start with small-scale pilot projects to evaluate and showcase the value of design, building momentum for broader adoption within the organisation. By addressing these barriers and fostering an environment that values and integrates design thinking, organisations can create a culture that takes a comprehensive approach to design, leveraging the efforts of multiple business units and fostering a cohesive and innovative work environment.

The chapter by Austin delves into the reality and complexity of the emergency department (ED) work system. Austin emphasises the need for a quality design strategy to make sense of the chaos within the ED and ensure efficient and safe patient care. This resonates deeply with my experiences in the health and safety field, where I've witnessed the immense pressure and demands placed on healthcare professionals in high-stress environments.

Austin advocates strongly for involving frontline workers in the design process. By engaging ED staff in identifying areas for improvement and co-creating solutions, organisations can tap into the invaluable knowledge and expertise of those directly involved in the work. This approach leads to more effective and efficient processes and enhances employee engagement and satisfaction.

To implement a successful quality design strategy in the ED, it's essential to break down silos and foster collaboration among different stakeholders, including doctors, nurses, administrators, and support staff. By promoting open communication, interdisciplinary teamwork, and a shared vision of patient-centred care, organisations can create an environment that supports continuous improvement and innovation.

Chari and Miller contribute a case study highlighting the importance of design partnership in healthcare. The HEAL (Healthcare Excellence AcceLerator) Bridge Lab exemplifies a design-led approach that prioritises user-centred, strengths-based, and innovative methods to improve the complex healthcare system. By involving consumers and clinicians, HEAL offers a constructive path towards healthcare transformation. This chapter highlights the complexity of good work design and emphasises the importance of exploration, experimentation, and adaptation in the face of change.

The concepts discussed in these chapters and the role of the CWDS underscore the importance of human-centricity in the workplace. They emphasise the significance of human relationships, integrity, and doing what is right. Amidst the competing demands of a workplace, we must remember that there is always a human at the end of it, benefiting from or driving the work. By embracing innovation, exchanging ideas, and daring to formalise these principles through roles like the CWDS, we can respond effectively to the future of work.

It is up to the trailblazers – the authors of this book, and readers like us – to make a compelling argument for the essential nature of work design skills in shaping the future. Let us forge ahead, leveraging the power of design to create better and more sustainable work and work environments.

21 The Restorative Potential of Older Australians Receiving Aged Care Services

A Matter of Policy and Practice

Jo Boylan

21.1 SERVICE DELIVERY MODELS

Despite having worked in aged care for more than 30 years, I am yet to see the adoption of a pervasive, relatable, sustainable national service delivery model that preserves the dignity of the aged care resident. I have chipped away at it, and I continue to try to advance this cause.

I will share my story:

> I am a woman in an executive management role of aged care services, and I have advocated extensively for an embedded healthy ageing approach across all services. The World Health Organization defines healthy ageing as 'the process of developing and maintaining the functional ability that enables well-being in older age' (World Health Organization, 2015). Regrettably, in 2023, older Australians continue to receive conventional aged care services that are not designed to meet their needs which could prevent or reverse circumstances that rapidly escalate into frailty, falls, dependency, disability, and their death (Okpalauwaekwe & Tzeng, 2021). There is considerable resistance by the sector and the Australian Government to constrain or cease traditional, medicalised, dependency-based service delivery. A positive shift would require a whole-of-system set of reforms with a transformational mindset that would include all involved in the system redesign. My story is one of leadership, challenge, and transfiguration in constraint-based models of care. I have worked to design systems and structures that help older Australians access the appropriate interventions to push back on their decline and disablement.

The World Health Organization (WHO, 2020) advances a global campaign on the reorientation of healthcare services towards healthy ageing, and this is urgent because

DOI: 10.1201/9781032711195-24

of the pace at which people are ageing. To achieve healthy ageing in an effective manner, those who regulate, deliver on, and reimburse this care must dismantle the outdated, conventional, narrow, and overtly medicalised systems in aged care and hospitals. Only then, can carers and their care recipients embrace effective, positive ageing models. This would require a whole-of-system redesign of service delivery that prioritises prevention, promotes health, builds capacity (systemically and among individuals), and values the creation of meaning in daily life. Such integration of care would recognise the potential of aged health consumers and their capacity to

- Learn new skills
- Find robust meaning in activities
- Foster new and existing social relationships
- Develop physical literacies and capabilities
- Improve their health profiles (e.g., through sound nutrition, activity, and routine, or through cognitive and psychosocial challenges to the 'just right' degree).

These new systems must enable consumers and carers to be adept at recognising the integrated risk factors and conditions that contribute to functional decline and understand the consequences arising from these factors. Action must be swift to intervene and mitigate frailty. If this occurred in Australia, it would make this country innovative in our approach to health, eradicating ageism by making 'healthy' normal for all, even the oldest Australian.

To enable this paradigm shift, identifying ageist attitudes and practices is a vital first step to improving healthcare outcomes for older people (Wyman et al., 2018). Without bias, aged care settings and hospitals can be set up with restorative care or rehabilitation as the initial default response to an older person's presenting needs. If done well, care recipients can be supported with a plan for services to assist them to regain their functional capacity and and/or fully recover from a setback.

This chapter is a call to action to appreciate the restorative potential of older adults. It also outlines why the Australian Government initiated a Royal Commission into aged care and examines the missed opportunity by the Royal Commission and the Australian Government to invest in authentic human rights and whole-of-system reforms to better deliver on this restorative potential. I have provided a three-step process for leaders to challenge the normalisation of ill being, explore the normalisation of well-being, and operationalise a model for healthy ageing in residential and home care services.

21.2 THE NEED TO INVEST IN QUALITY AND SAFETY IN AGED CARE

The Australian Government and aged care providers have historically been inattentive to the drivers of disability and have been blind to the restorative potential of older adults (Rudnicka et al., 2020). Ibrahim (Royal Commission into Aged Care Quality and Safety, 2020) states that the failure to act (or to decide on appropriate action) and the failure to invest in quality and safety are not the sole responsibility of the aged care provider but the aged care system – this includes government policymakers and regulators. At the point of care, this involves the aged care provider, its governance, the operational managers, trainers, and clinical and support staff to change negative attitudes on ageing and promote functional capacity. Beard and Bloom (2015) any argued this case in the First World Report on Ageing and Health (WHO, 2015). This report recognised that the health of older people remained dominated by rapid decline in physical capacity in later years and subsequent reduced quality of life. Functional decline is more common when a person is unable to engage in everyday activities, influenced by their age, lifestyle, multiple health conditions, and cognitive functioning. This decline can cause a loss of independence, disability, and lead to further health complications (Covinsky et al., 2003), which are costly to the public purse.

It is compelling to believe that the Australian Government and the aged care industry can make better decisions and authentically invest in enabling older Australians to remain a resource to their families, communities, and economies when they thrive in their later years. However, to enable these changes, we must understand the contributing factors that cause healthy ageing. This must become a reality for older people at the heart of where they live (including in residential care and home care). Researchers have urged the government, as a public health imperative, to invest in reorienting services and settings towards healthy ageing in preparation for the dramatic increase in numbers of older Australians in the future and the dramatic decline in skills and staff (Dawes & Topp, 2022; OECD, 2009).

Although there is extensive evidence for the Australian Government to reinvestment and reorientation of policies towards healthy ageing, there is no comprehensive restorative regulation or funding model to support this. The government has not systematically dismantled the barriers to the structures and social systems that constrain older people in their pursuit of health or recovery from setbacks. There is no policy direction, regulation, or funding to guide aged care providers in eradicating ageism, disparity, or inequality in a meaningful way. This would enable older Australians to access necessary care and support that will assist them to regain their functional capacity, fully recover from a setback, and enable a positive trajectory towards restored health. For example, to access rehabilitation care, government funding must adequately cover this care by the providers. We must also help our aged family and friends enjoy the dignity of a good death, which only comes from investing in their health right to the edge of their lives.

The WHO (2015) report argues that to achieve the goal of healthy ageing, there must be a reconstituted integrated, whole-of-system and service approach that centres on the needs and rights of older people. The report argues the need to deliver care around a common goal of functional ability to enable connections to wellbeing. Timely access to exercise shows that even frail elderly people can experience improved physical health (Wise & Nutbeam, 2007). One might form the logical hypothesis that health promotion should be core business in aged care (as it has been in the facilities that I manage).

21.3 CALL FOR A ROYAL COMMISSION INTO AGED CARE

Between 2019 and 2021, the Australian Government initiated a Royal Commission into Aged Care to understand the care provided to older people and to evaluate what needed to change to improve aged care service delivery. During the public hearings, workshops, and submissions, the Commissioners heard from Australians who shared their stories of neglect and exploitation. In the Aged Care Quality and Safety (2020), it described the Australian aged care system as inadequate because it had not been prioritised by successive governments. It also found low financial investment by the Australian Government into the whole aged care sector (Dyer et al., 2020).

The Royal Commission was born out of the 'Oakden Mental Health Services for Older Persons' tragedy' (Aged Care Quality and Safety, 2021). The Oakden facility shut down in 2017 but came to symbolise wider concerns for Australia's aged care sector. Investigation by South Australia's anti-corruption body and Independent Commissioner Against Corruption (ICAC) between 2017 and 2018 found that Oakden consumers lived in a facility which could only be described as a disgrace, and in which they received extremely poor care (Royal Commission into Aged Care Quality and Safety Interim Report, 2019b). The absurdity of these findings is that Oakden was a State Government–run facility but funded and audited by the Commonwealth Government.

A further paradox is that between 2006 and 2009, I was part of a group of practitioners who were engaged by the State and Commonwealth Government (for three years) to go into Oakden and 'fix' the 26 non-compliances Oakden had received in an accreditation visit. The findings of the audit emphasised a culture of poor-quality care and poor outcomes for their consumers. This was despite being staffed at extraordinary levels by trained nurses.

During the first couple of months, I cried most days after work while on my way home. I felt turmoil and distress at seeing the level of care provided to consumers: it was a breach of human rights. Several months later, I stopped crying: I watched the culture of care improve as consumers were treated with more dignity and humanity because of our oversight and the framework of person-centred care that was instituted.

When hospitals or aged care homes develop a poor culture exacerbated by poor leadership, the normalisation of deviance is enabled (Barach & Phelps, 2013) performance drifts from the expected performance quality norms, and poor practice 'starts to feel right' or, at least, business as usual. The healthcare workers at Oakden had become insensitive to these deviant practices and subsequently the older people experienced episodic neglect, mostly through the provision of substandard care. In 2009, our contract was completed and Oakden remained compliant for a number of years before it fell back into a culture of poor quality care.

21.4 RESILIENCE DURING THE ROYAL COMMISSION INQUIRIES

During the Royal Commissions inquiry, I was employed as the Executive of Services for a well-regarded and trusted not-for-profit organisation (NGO), providing leadership and governance across a large residential and community portfolio. The NGO's Board, CEO, and executive team were committed to and invested in a healthy ageing approach: They were determined to provide a high standard of service delivery. In this role (and in other previous senior roles), I worked a minimum of 10–12+ hours per day and most weekends to ensure the implementation of an elevated level of clinical governance and service delivery. This level of monitoring, governance and excessive paperwork meant that I was a workhorse (commonly termed 'a workaholic').

Our organisation invested in fitness facilities in all our homes, though it was more than a case of 'build it and they will come': we focused on interventions and practices aimed at preventing avoidable decline and promoting healthy ageing. Subsequently, it had a large interdisciplinary health-professional-led model driving social and physical engagement. Our organisation was one of a minority that embedded a healthy ageing approach into service delivery, and we provided the continuum of care through robust residential and community services with an extensive nursing and allied health team. We were intent on increasing consumers' quality of life: we strove to keep people walking and we enabled them to live well until they died.

The practice of promoting healthy ageing and the provision of sound clinical governance go hand in hand. When a staff member started work at our organisation, we expected that they provide all consumers and clients with equal access to all known interventions that would improve their life. Our staff were part of what we viewed was essential to place consumers on a positive trajectory towards a good life and, eventually, a good death. The expectations set by our organisation meant that there was extraordinary multidisciplinary team work to problem solve the pathophysiology, psychosocial, and musculoskeletal issues preventing a resident or client from progressing along a healthy ageing path.

Healthy ageing practices are foreign to aged care, yet this organisation, through strong leadership, committed to a restorative direction across all practices and programmes and we fought to make this the 'new normal'. I encouraged us all to push back against any deviation from this new norm and to contest any lower standards of care. Our goal was to promote the psychosocial and physical capacity of the people whom we served.

21.5 STORIES OF NEGLECT AND EXPLOITATION DURING THE ROYAL COMMISSION

It is difficult to describe the enormous workload that came with the commencement of the Royal Commission and their initial request for aged care organisations to self-declare 'incidents or complaints of substandard care' from the last five years (December 2013 to December 2019). The Commission informed aged care organisations that failure to do this may result in closer scrutiny. They advised aged care organisations that when there were instances or complaints of substandard care, mistreatment, or abuse, it would be vital that they engage an experienced external team (solicitors and counsel) for support. This information had to be prepared and delivered in a compressed time over the Christmas period in 2019 by the organisation's internal team before submission to the external team of solicitors and counsel. We invested insurmountable overtime and additional resources to examine every complaint and incident that could be classified as 'substandard care'. I found complaints and incidents that arose from substandard care, poor practice, or inaction; however, this was not pervasive. Our teams routinely identified, investigated, and remedied these events, so they were infrequent during this five-year reporting period. Because the organisation's multidisciplinary team had already adopted a healthy ageing mindset and were prepared to implement timely health-oriented responses, the interventions revealed extraordinary examples of recovery from setbacks. There was hard evidence of our systematic efforts that caused health improvements among aged care consumers even when operating under the constraints of poor government funding and burdensome regulations.

In October 2019, the Royal Commission into Aged Care Quality and Safety's Interim Report found that the aged care sector failed to meet the needs of its older, vulnerable Australians. Commissioners Richard Tracey AM, RFD, QC, and Lynelle Briggs's AO investigation led them to describe the aged care system as one of 'neglect' (Royal Commission into Aged Care Quality and Safety, 2019b). During this time, my workload averaged 12–14 hours most days and included working most weekends. It took all my energy to persevere and not give up because leading our teams through this heavy claim of neglect and substandard care was so difficult.

Mike Baird, the former Liberal premier of NSW, explained in an Opinion Article (Baird, 2021) that he also felt this enormous sadness reading the first Royal Commission Report. The examples of neglect wounded deeply. He reported feeling extreme sadness as the adverse events seemed inevitable given the lack of priority funding, governance, and resources that the sector suffered in cumulative effect over decades. Baird (2021) explained,

> The Royal Commission has a damning view on how we got here with an approach described as 'financial'. It has felt like the Government's main consideration was what was the minimum (financial) commitment it could get away with, rather than what should be done.
>
> **(p. 1)**

During the Royal Commission, the Commissioners asked 'how' and 'why'. They questioned the decisions that the government, providers, educators, and researchers made. There were hearings, workshops, and submissions of disempowerment, ageism, inadequate staffing, inferior quality, and costs of aged care. In a witness statement by Professor Joseph Ibrahim, he intimated that 'Residential Aged Care homes are a place where frail older people receive basic health and other care, while staff, health professionals, and family may wait for them to die' Royal Commission into Aged Care Quality and Safety, 2020, p. 1). Ibrahim (Royal Commission into Aged Care Quality and Safety, 2020) emphasised that the onus of responsibility was on everyone to ensure that consumers receiving aged care services thrive.

21.6 ADDRESSING THE ROYAL COMMISSION ON A HEALTHY AGEING APPROACH

The Royal Commission invited me to speak about the use of allied health professionals in Australian aged care services. The Commissioners and lawyers interviewed me about these issues. I provided a witness statement in advance with evidence about the benefits of normalising healthy ageing for older Australians receiving aged care services using a multidisciplinary approach.

They asked me to evaluate how I thought that the organisation that I led was using allied health services for consumers in each of the residential care homes. I explained to the Commissioner that our organisation had designed gyms and provided exercise rehabilitation service as essential resources in each of the homes and community service centres. This resident's active participation in meaningful activity complemented these services. The multidisciplinary teams – nurses and allied health staff working together – operated the programmes with the unified goal of reversing frailty by enabling older people to become stronger and resilient. Subsequently, they stayed above the disability line. I explained that we had implemented the required staffing and structures. I acknowledged that we had an exceptional board and team of executives that were committed to the cause. They understood their responsibility for shaping 'well-beings' and not 'ill-beings'. I explained that although the Government funded deficit models of care, e.g., Aged Care Funding Instrument (ACFI) and Aged Care Classification funding model (AN-ACC), our organisational leaders had a strong commitment to health promotion, and it was central to the way that we did business. We understood that quality of life feedback from consumers and their families was more positive in a healthy ageing model, and we appreciated receiving this feedback:

> By leveraging each of the disciplines strengths the collective goal was to prevent avoidable decline and push back on disability for each resident or client, thereby extending opportunities for recovery every time they fell below the disability threshold. This is a human right.
>
> **(Jo Boylan, Royal Commission Statement, 2019a)**

I explained to the Commissioner that we achieved amazing outcomes with an embedded healthy ageing approach. This included significantly reduced wounds, reduced hospital transfers, and increased functionality in consumers when contrasted with industry benchmarks. I explained that over 90% of our consumers were walking until the day of their death (healthful to the end). I explained that we implemented a multidisciplinary team approach (combining the work of nursing and allied health teams) to monitor and care for our consumers. We used systems to detect anyone drifting towards the disability threshold and we had to act quickly once we noted a resident in decline. A skilled team used appreciative inquiry techniques (Hung et al., 2018). We leveraged what was important to people to prepare their recovery pathway. Once recovery needs were detected, we placed consumers and clients on a 12-week intensive recovery plan. Often, consumers with good muscle memory finished their recovery plan at week five because they met their goals early. I explained to the Commissioner that giving older people this opportunity to recover from setbacks or to reverse their level of frailty by accessing recovery programmes was a human right to preserve basic dignity. Our organisation understood 'the dance of responsibility and leadership' and the power to create well-being by pushing back on disability. 'Accessible early intervention and restorative services should be readily available each time a person falls below the disability threshold. The first defeat is not the last defeat', Jo Boylan, Royal Commission Statement (2019a).

21.7 ERADICATING AGEISM AND NORMALISING HEALTHY AGEING

I am certain that recognising the normalisation of deviance (when low standards become accepted) and, conversely, knowing the indicators of healthy ageing for older

Australians supports the fundamentals of quality care. A mindset for helping all older people to thrive, without discernment or discrimination, is missing in aged care but this can change. We must eradicate ageism and then reengineer the rationalised resources. We must use the power of storytelling to translate healthy ageing and we must innovate our care methods.

The WHO (2020) has increased its campaign to eradicate ageism and discrimination, viewing it as a systemic form of oppression and emphasising that there is nothing more dangerous than ageism for older people receiving healthcare. In 2023, older Australians remain at risk of health disparity (the difference that affects a person's ability to achieve their best health) and inequality (reduced opportunities). To eradicate ageism, we need to instil a 'growth mindset' for positive and healthy ageing (Chasteen et al., 2022; DeBacker et al., 2018).

Carol Dweck, the psychologist who introduced the world to growth and fixed mindsets, studied the differences of these orientations (Dweck, 2010). She believed that growth mindset achievements are reflective of the effort to learn new things while accepting it is possible to enhance intelligence and ability. When aged care training is reoriented towards cultivating a growth mindset and towards eradicating ageism, aged care will provide the highest attainable standard of health, which is the World Health Organisation (WHO, 2020) views as a fundamental right of every human being. A 'healthy ageing' mindset and practice in aged care is fundamental if we are to enable well-being among older people (WHO, 2020). Comprehensive training around developing a health promoting mindset is as important and necessary for the healthcare worker as it is for an older person receiving their aged care services. Together, we can achieve successful ageing. 'The requirements (systems and structures) of prevention and restorative care for older Australians needs to become a core component of the curriculum for all health-related VET & tertiary degrees' (Royal Commission Statement, 2019a).

21.8 HEALTH PROMOTION AS CORE BUSINESS IN AGED CARE

During the Royal Commission, the Commissioners reviewed innovative models of aged care in Australia. This review was prepared by the Commissioners in November 2019. The review identified that aged care services had not embedded rehabilitation, reablement, and restorative approaches across Australia. The review highlighted that the funding instrument (ACFI) for residential care homes was based on dependency and with limited incentives or funding to increase resources to achieve recovery when a resident returned from hospital or when functional decline occurred.

The findings identified that other countries receive government funding to embed health promotion as core business. Yet, following this review, the Australian Government implemented a new funding tool, AN-ACC in 2021, that rewarded even greater levels of consumer dependency. More astonishingly, the Royal Commission recommendations aligned the new funding tool to care minutes, funding nurses and care workers only to provide care. These minimum care minute responsibilities align to a case mix adjusted funding model (case mix used in hospitals) with no inclusion of rehabilitation, restoration, or reablement to fund the allied health practitioners. The adage, 'what you fund is what you get' is now the reality for many older Australians receiving aged care services. It will take the commitment from providers and practitioners to work out ways to bring health promotion as core business through the nursing and care team as a constraint-based innovation.

21.9 DISMANTLING THE BARRIERS TO HEALTHY AGEING FOR AUSTRALIANS RECEIVING AGED CARE SERVICES

For the last 20 years, I have practised and led authentically from a 'healthy settings approach'. A healthy settings approach has its roots in the WHO 'Health for All' strategy laid out in the 1986 Ottawa Charter for Health Promotion. It is a whole-of-system approach in health promotion and prevention and integrates holistic and multi-disciplinary approaches across risk factors.

The 'settings' referred to by this approach are residential homes and home care services. Older Australians receiving services are on average 87 years old, with risk factors that include multiple comorbidities (between 8 and 15) compounded by frailty, poor function, reduced social engagement, and lack of access to services (Commission, Productivity, Caring for Older Australians, 2011) that will assist them to regain their functional capacity (Thapaliya et al., 2023).

Aged care in Australia delivers basic and traditional care. Traditional methods of service delivery include personal care assistance with showering/bathing, dressing, mobility and transfers, toileting, food/meal assistance, cleaning, and social support. Due to the older person's increased dependency, the service is 'done to them' rather than in a re-enabling way or with a plan for services to assist them to regain their functional capacity and or fully recover from setbacks. The aged care setting is not set up or funded to ensure that restorative care or rehabilitation services are available to meet an older person's needs. In the past two decades, there have been great strides to advance older peoples' rights; however, we have not addressed the inequity or disparity for older people to access embedded health promotion in their aged care services. With the changes to funding (AN-ACC and care minutes), Australian providers will need to demand a comprehensive public health response to developing policies that can secure resources to provide for essential well-being in older age.

A more comprehensive understanding of healthy ageing in aged care starts with research. Improving the collection, integration, and synthesis of relevant data and information into everyday care practices must be a priority for providers. This will help inform meaningful decision-making about service impact on progressing healthy ageing for older Australians. Policymakers must influence the funding available to make healthy ageing normal.

21.10 A MODEL FOR HEALTHY AGEING

Actions to promote healthy ageing of older people through different settings and multidisciplinary methods requires fundamental change at every level. The expectations, leadership, growth mindset, and belief that healthy ageing is a human right must start at the top, with the Board and the Executives.

To help older Australians thrive and to prevent their decline while receiving aged care services, I have designed a model to progress healthy ageing using three critical steps:

1. To shift our thinking about age and aged care – provide workshops
2. To understand the theory of healthy ageing, best practice standards and coordination, and health promotion principles
3. To develop and implement a model of healthy ageing

The three exemplars are brief resources for aged care homes and services to guides and drive a whole-of-system response which places health promotion and healthy ageing as core business to service delivery.

In summary, Australia has a great opportunity to take the lead in addressing the social determinants of healthy ageing, including eradicating ageism and providing our older generation with 'early intervention' and 'prevention' to keep them functionally well and above the disability threshold right up until they die. This transformative and positive change creates great outcomes for older people, preserving and protecting their dignity, respect, and human rights (WHO, 2020).

In Exemplar 1, have outlined the three fundamental workshops that must occur to shift aged care and healthcare workers' thinking and prepare for a healthy ageing model.

Exemplar 1

1. The first step requires examining our own stereotyping, prejudices, and discrimination. The workshop must be designed to challenge how we
   a. think about age and aged care;
   b. recognise and understand the negative effects of our attitudes, on our practices, and potential impact on consumers; and
   c. use resources that create dependency and even cause 'ill-being' rather than create 'well-being'.
2. In these workshops, we discuss the following:
   a. Why there is greater investment in basic and traditional sedentary care rather than restorative care and rehabilitation (accepting one's chronic conditions rather than assuming that these are barriers to participation)?
3. We review the investment for embedded restorative care and rehabilitation, and ask:
   a. Which people should drive restorative care and rehabilitation?
   b. What structures are needed?
   c. What systems will need to be developed and implemented?
   d. Where is the point of access to restorative care and rehabilitation?
   e. Who and how will we recognise frailty and decline and increasing dependency?
   f. Who and how will we implement early intervention and prevention pathways (intensive 12-week pathway to recovery and then onto a maintenance pathway)?

   g. What resources are needed?
   h. What will need to happen to achieve equity in access to health-promoting resources for all consumers?
4. Finally, the group will acknowledge and pledge to the changes and enormous effort and grit that is required for a complete system paradigm shift towards a healthy ageing and human rights model. Once everyone has agreed to invest in healthy ageing and improved outcomes for older people, the environment, systems, and structures can be redesigned.

In Exemplar 2, I have clearly defined and articulated the model of healthy ageing, core elements, and principles to help ensure that all aged care workers are working towards common goals: a person-centred and comprehensive set of services to prevent, slow, or reverse decline, including evaluating their performance.

In Exemplar 3, I have outlined the key stages for developing and implementing a model for healthy ageing: (1) planning stage, (2) development stage, (3) implementation stage, and (4) evaluation stage.

TABLE 21.1
Exemplar 2

Exemplar 2	Defining the Model of Healthy Ageing
Defining the model of healthy ageing	This organisation will agree to commit to implement an integrated, whole-of-system and service approach that is centred around the needs and rights of older people to access restorative care as a preventative strategy and/or a plan for services to assist them to regain their functional capacity and/or fully recover from setbacks
Key objectives and goals	All services will deliver care around a common goal of functional ability to enable older people to achieve what they most value
Core elements	Overview
The theory of healthy ageing	The World Health Organization defines healthy ageing as 'the process of developing and maintaining the functional ability that enables wellbeing in older age' (WHO, 2015)
Principles of best practice care underpinning the clinical care standards	Key elements of best health-promoting care coordination (multidisciplinary team approach) include the following: • A person-centred approach • Treating people with dignity and respect • Communicating with each person (or key representative) about their clinical needs and access to health-promoting interventions to improve their health outcomes • Encouraging participation in decision-making using appreciative inquiry techniques (health literacy) – having caring conversations about what is working well, understanding why these aspects work well, and co-creating strategies to help these good practices to happen more of the time

TABLE 21.1 *(Continued)*
Exemplar 2

Exemplar 2	Defining the Model of Healthy Ageing
Five principles of health promotion (adapted to aged care context) (WHO's Ottawa Charter 1986)	• Develop consumers and workforce personal and professional skills to enable people to take control over the determinants of their health (or facilitate this process) and thereby improve their health outcomes • Create health promoting environments, including holistic and multidisciplinary methods which integrate actions across risk factors (maximising prevention) • Embed organisational action to progress healthy ageing • Develop, implement, and monitor healthy ageing policies/procedures/performance indicators • Reorient aged care services core business to a model of healthy ageing

TABLE 21.2
Exemplar 3

Exemplar 3	Four Key Stages in Developing and Implementing a Model of Healthy Ageing
1. Planning stage	Overview
Scoping the problems and issues across aged care services and summarise the current model of care	Traditional, medical, dependency-inducing model is currently the aged care sector's default position for older frail people who have restorative/rehabilitation needs and can recover from setbacks given access to early interventions that will reverse frailty, reduce dependency/disability, and place them on a positive trajectory • Potential for ageist, discriminative attitudes, fixed mindset thinking with entrenched policies and practices that limit opportunities to access restorative care and rehabilitation care as needed • The 'setting' (structures and systems) at risk of creating 'ill-beings' instead of 'well-beings'
Establishing baseline data	• Measure % of mobile consumers and current engagement in early intervention recovery programmes and their improvement and % engaged in consistent social activities, and measure their quality of life
Identify factors to optimise sustainability	• Map and plan what is needed – environment (gyms, safe measured walking areas), structures (fitness leaders), systems (early intervention procedures), processes (recovery programme/monitored and evaluated) – and indicate who is responsible. Develop Gantt chart/project plan

(Continued)

TABLE 21.2 *(Continued)*
Exemplar 3

Exemplar 3	Four Key Stages in Developing and Implementing a Model of Healthy Ageing
Identify key stakeholder involvement, key leaders	• The Board and CEO and key Managers must endorse the model for healthy ageing and the multidisciplinary team must implement this – this includes the fitness leaders, exercise physiologists, other allied health professionals, nurses, lifestyle coordinators, and care workers
2. Development stage	Overview
Streamlining and standardising the processes	• Identify and document responsibilities and processes for recognising frailty/decline (screen/report), early intervention pathways, maintenance pathways, multidisciplinary meetings (weekly) • Develop ongoing workshops/educate teams about restorative and rehabilitative processes/skill sets/develop champions and highlight positive case studies and achievements
Establishing data management systems	• Tools for measuring (short physical performance battery test, barthel index, berg balance, grip strength), and how to integrate data gathering into day-to-day practices • Establish who will measure, monitor, evaluate, and report through to the Board
Establishing performance indicators	• % of consumers participating in physical and social activities and individual improvement outcomes from early intervention pathways and review trajectory of how consumers die and success in compression of morbidity (dwindling versus walking until they die)
Establish monitoring systems to ensure changes lead to improvements	• Establish a steering committee to ensure delivery on the model • Develop comprehensive procedures to guide all change processes and set up a structure to audit, evaluate and meet key performance indicators
Skill development	• Conduct ongoing workshops, education, and training in appreciative inquiry, fixed mindset versus growth mindset, ageism, normalisation of deviance, early identification of frailty/decline and early intervention pathways to treat frailty/decline, and maintenance pathways to prevent setbacks • Review/change all role descriptions to outline expectations for health promotion, best practice coordination, and healthy ageing
Pilot testing of model case studies to build the team spirit	• Create success stories about consumers who have improved through the collaborative efforts of multidisciplinary teamwork and consumer commitment. Place these stories up in the gyms and in newsletters

TABLE 21.2 *(Continued)*
Exemplar 3

Exemplar 3	Four Key Stages in Developing and Implementing a Model of Healthy Ageing
3. Implementation stage	Overview
Support of key leaders and clinical team	• Do not deviate from your healthy ageing message. Have grand expectations and promote this conviction with all communication (from the Board, CEO, and across all key leaders)
Communication strategy	• Make all messaging/stories compelling and craft powerful narratives/pitches for why healthy ageing matters • Ensure it is clear and personalised
Daily leadership and check-ins	• Key leaders must check in daily that all steps and stages are implemented successfully
4. Evaluation stage	Overview
Measuring performance against pre-specified performance indicators	• Establish baseline performance indicators • Ensure that the success indicators align with the goals and activities, and that there are tools/systems for accurate, timely measuring, recording, and reporting of data
Evaluating what needs adjustment/ rethinking	• Through the steering committee/group: re-evaluate and adjust to best practice
Impact of change on people and systems	• Document/report (case studies/set measures) the impact of changes on people to assist with evaluation and adjustments as needed

These stages are fundamental in creating the structures, systems, and setting for healthy ageing to improve the lives of older people and the services that they receive. Fundamental shifts will take place, not only in actions but in how people think about ageing going forward (WHO, 2020).

REFERENCES

Baird, M. (2021, March 24). How to repair aged care and heal the nation's broken heart (opinion article). *The Sydney Morning Herald*. https://www.smh.com.au/national/how-to-repair-aged-care-and-heal-the-nation-s-broken-heart-20210323-p57d9e.html

Barach, P., & Phelps, G. (2013). Clinical sensemaking: A systematic approach to reduce the impact of normalised deviance in the medical profession. *Journal of the Royal Society of Medicine*, *106*(10), 387–390. https://doi.org/10.1177/0141076813505045

Beard, J. R., & Bloom, D. E. (2015, February 14). Towards a comprehensive public health response to population ageing. *Lancet*, *385*(9968), 658–661. https://doi.org/10.1016/S0140-6736(14)61461-6. Epub November 6, 2014. PMID: 25468151; PMCID: PMC4663973.

Chasteen, A. L., Schiralli, J. E., Le Forestier, J. M., & Erentzen, C. (2022). Age stereotypes and ageism as facets of subjective aging. In *Subjective views of aging: Theory, research, and practice* (pp. 229–247). Springer International Publishing. https://chasteen.psych.utoronto.ca/publications

Commission, Productivity, Caring for Older Australians. (June 28, 2011). Retrieved, from https://ssrn.com/abstract=2006095 or http://dx.doi.org/10.2139/ssrn.2006095

Covinsky, K. E., Palmer, R. M., Fortinsky, R. H., Counsell, S. R., Stewart, A. L., Kresevic, D., Burant, C. J., & Landefeld, C. S. (2003, April). Loss of independence in activities of daily living in older adults hospitalized with medical illnesses: Increased vulnerability with age. *Journal of the American Geriatrics Society*, *51*(4), 451–458. https://doi.org/10.1046/j.1532-5415.2003.51152.x. PMID: 12657063.

Dawes, N., & Topp, S. M. (2022). Challenges to managing quality of care in northern Queensland residential aged care facilities. *Social Sciences & Humanities Open*, *6*(1), 100300. https://doi.org/10.1016/j.ssaho.2022.100300

DeBacker, T. K., Heddy, B. C., Kershen, J. L., Crowson, H. M., Looney, K., & Goldman, J. A. (2018). Effects of a one-shot growth mindset intervention on beliefs about intelligence and achievement goals. *Educational Psychology*, *38*(6), 711–733. https://doi.org/10.1080/01443410.2018.1426833

Dweck, C. S. (2010). Mind-sets. *Principal Leadership*, *10*(5), 26–29. www.mindsetworks.com/science

Dyer, S. M., Valeri, M., Arora, N., Tilden, D., & Crotty, M. (2020). Is Australia over-reliant on residential aged care to support our older population? *Medical Journal of Australia*, *213*(4), 156–157, e1. https://doi.org/10.5694/mja2.50670

Hung, L., Phinney, A., Chaudhury, H., Rodney, P., Tabamo, J., & Bohl, D. (2018). Appreciative Inquiry. *International Journal of Qualitative Methods*, *17*(1). https://doi.org/10.1177/1609406918769444

OECD. (2009). *Ageing workforce* (pp. 72–73). OECD. https://doi.org/10.1787/9789264061651-16-en

Okpalauwaekwe, U., & Tzeng, H.-M. (2021). Adverse events and their contributors among older adults during skilled nursing stays for rehabilitation: A scoping review. *Patient Related Outcome Measures*, *12*, 323–337. https://doi.org/10.2147/prom.s336784

Ottawa Conference Report. (1986). *Reorienting health services, health promotion 1.4* (pp. 459–460). www.who.int/teams/health-promotion/enhanced-wellbeing/first-global-conference/emblem

Royal Commission into Aged Care Quality and Safety. (2019a). *Statement of Josephine Boylan*. https://agedcare.royalcommission.gov.au/sites/default/files/2021–03/final-report-volume-2_0.pdf

Royal Commission into Aged Care Quality and Safety. (2019b, October 31). *Aged care in Australia: A shocking tale of neglect*. The Royal Commission into Aged Care Quality and Safety Interim Report Released. Retrieved June 4, 2023, from https://agedcare.royalcommission.gov.au/news-and-media/royal-commission-aged-care-quality-and-safety-interim-report-released

Royal Commission into Aged Care Quality and Safety. (2020). *Statement of Professor Joseph Elias Ibrahim)*. https://agedcare.royalcommission.gov.au/system /files/2020–06/WIT.0115.0001.0001.pdf

Royal Commission into Aged Care Quality and Safety, Final Report; Care, Dignity and Respect (Released). (2021, March 1). https://agedcare.royalcommission.gov.au/sites/default/files/2021–03/final-report-volume-3a_0.pdf

Rudnicka, E., Napierała, P., Podfigurna, A., Męczekalski, B., Smolarczyk, R., & Grymowicz, M. (2020 September). The World Health Organization (WHO) approach to healthy ageing. *Maturitas*, *139*, 6–11. https://doi.org/10.1016/j.maturitas.2020.05.018. EPUB: 2020, May 26. PMID: 32747042; PMCID: PMC7250103.

Ryburn, B., Wells, Y., & Foreman, P. (2009, May). Enabling independence: Restorative approaches to home care provision for frail older adults. *Health Social Care Community*, *7*(3), 225–34. https://doi.org/10.1111/j.1365-2524.2008.00809.x. EPUB: 2009, January 20; PMID: 19175428.

Thapaliya, K., Cornell, V., Lang, C., Caughey, G. E., Barker, A., Evans, K., Whitehead, C., Wesselingh, S. L., & Inacio, M. C. (2023). Aged and health care service utilization by older Australians receiving home care packages. *Journal of the American Medical Directors Association, 24*(3), 395–399, e2. https://doi.org/10.1016/ j.jamda.2022.11.019

Wise, M., & Nutbeam, D. (2007). Enabling health systems transformation: What progress has been made in re-orienting health services? *Promotion Education, 14*(Suppl 2), 23–27. https://doi.org/10.1177/10253823070140020801x. PMID: 17685076.

World Health Organisation (WHO). (2015). *World report on ageing and health.* World Health Organization. https://apps.who.int/iris/handle/10665/186463

World Health Organisation (WHO). (2020). *United nations decade of healthy ageing (2021–2030).* World Health Organization. www.who.int/initiatives/decade-of-healthy-ageing

World Health Organisation (WHO), Regional Office for Europe. (1986). *Ottawa charter for health promotion.* World Health Organization, Regional Office for Europe. https://apps.who.int/iris/handle/10665/349652

Wyman, M. F., Shiovitz-Ezra, S., & Bengel, J. (2018). Ageism in the health care system: Providers, patients, and systems. *Contemporary Perspectives on Ageism*, 193–212. https://link.springer.com/chapter/10.1007/978-3-319-73820-8_13

22 Architecture and Salutogenesis

Jan A. Golembiewski and Narges Farahnak Majd

22.1 INTRODUCTION

Throughout the last 20 years, the effect of the built environment on well-being has become a topic of interest and is increasingly posited as a factor for social sustainability (Olakitan Atanda, 2019). Healthy architecture is also seen for its potential economic upside, especially when it comes to health architecture (Pentecost, 2013) and workplace design (Hedge, 2017). Despite the recent noticeable shift towards well-being in design, the quality of the architectural milieu and how it meets human needs are not new concerns. Since people started building, other disciplines have informed architecture to make it better fit its users. For millennia, these efforts have relied on trial and error, religious doctrine, art practice, instinct, and superstition. Since the 1980s, the effort has exploded, with new insights from political humanism (feminism, post-colonialism, etc.), the hard sciences (neuroscience, chemistry, biology, etc.) and the behavioural sciences (psychology, sociology, anthropology, marketing, etc.). How architecture shapes human desires, behaviours, thoughts, emotions, and mental well-being can now be predicted better than ever before – it is even being used (as Canadian architect, Tye Farrow tells us) 'to create health' (Farrow, 2019). Much of this work does not involve architecture, but what is outside of it: views into the natural environment (Ulrich, 1979, 1984) and the closely related Attention Restoration Theory (Kaplan & Kaplan, 1989, 1982). But assuming that architects are either willing to provide views of nature or that they simply cannot for reasons that might be beyond their control (like the site a client chooses, for example), there is still plenty more the architect can do to improve health through design.

Designers of the built environment have traditionally focused on health factors that reduce the negative influences on comfort and safety (noise, and air pollution, organic and chemical toxins, sensory overloads etc. [e.g., Halpern, 1995]). But increasingly, psychological and sociological responses to architectural decisions have gained prominence. This set raises the complex interrelations of the human and the built environment to the forefront as it explores the potential of the environment to foster positive emotions, well-being, and engagement (Xie et al., 2017). In this regard, the 'Salutogenic Method' (Golembiewski, 2010) is possibly the most thorough methodology designers can employ to ensure that architecture is doing its part, so buildings' users live life to the fullest and healthiest. The salutogenic method is a practical application of the theory of Salutogenesis formulated by Antonovsky (1979). The model seeks to improve health by identifying health-promoting factors

 DOI: 10.1201/9781032711195-25

rather than eliminate the factors that undermine well-being. It examines what shapes our capacity to adapt to stressors in the environment such as life challenges and diseases, often viewed as inevitable.

Since Alan Dilani first introduced Antonovsky's theory into architecture (Dilani, 2001), researchers and architects have used the model to advance well-being through design: in workplace design (Ruohomäki et al., 2015; Roskams & Haynes, 2020); in healthcare facilities design such as in hospitals (Dilani & Armstrong, 2008; Golembiewski, 2017a; Abdelaal & Soebarto, 2019), psychiatric facilities (Golembiewski, 2010), residential aged care centres and centres for people with dementia (Chrysikou et al., 2018; Golembiewski, 2017b), and in the public realm and urban scale like communities, neighbourhoods, town, and cities (Vaandrager & Kennedy, 2017; Maass et al., 2017).

Salutogenesis is a powerful tool, but marketers may use the term to label projects that offer little by way of salutogenic value. As an example, the project of Lady Cilento Children's Hospital (by Conrad Gargett Riddel and Lyons Architects) struggles to live up to the salutogenic principles the scheme is marketed with (Lyon, 2017); the hospital has a deep floor plate, and even the double-height spaces radiating out from two vertical atriums do not allow much sunlight penetration. Furthermore, many interior rooms also have no external windows to allow even views of the city, much less views of nature. As we will see, all are elements that support a higher sense of coherence – the foundation of the salutogenic principle that supports health. Stephanie Brick is another example of an architect who has also picked up 'salutogenic' as a buzzword, even though it's difficult to identify any features of her work that appear to be any more salutogenic than any other designers of the same typology (Brick, 2020).

Salutogenesis is a broad, robust, and comprehensive theory to predict the circumstances that promote well-being regardless of typology. But hitherto, writing about salutogenesis in architecture is most often used as a proxy for the Biophilia Hypothesis and Attention Restoration Theory (e.g., Abdelaal & Soebarto, 2019); on specific typologies (e.g., Christikou, et al., 2017; Golembiewski, 2017b, which are both exclusively concerned with dementia care; Golembiewski, 2010, with mental healthcare, etc.); or are simply vague about what salutogenesis is and how to apply it (e.g., Dilani, 2001, 2005). In this chapter, we look at salutogenic architecture, and identify salutogenic approaches in architectural projects far more broadly, yet still with specificity. The selected projects help indicate how the translation of salutogenic principles into architecture might appear, and the features we discuss are interpretative rather than prescriptive. The selection was based on the authors' expertise in architecture and salutogenic theory and does not assume the architect knew of salutogenic theory. This chapter will help designers, architects, stakeholders, and planners who want to maintain and improve health among occupants, who value human needs, and who wish to enable the natural recovery process for healthcare recipients.

22.2 SALUTOGENESIS: WHAT IS IT?

Salutogenesis is a theory of health that presents an alternative perspective to the normative pathogenic model, which focuses on the origins of individual ailments and diseases to remove them, protect against them, or repair the damage they have done. Instead, the salutogenic model regards health as a continuum with the abstract concepts of death and total engagement in life as its bookends. The model revolves around a central construct called the 'sense of coherence' (SOC), which marks where a person stands along the 'healthy to dis-ease continuum' (Antonovsky, 1987). Salutogenesis supports the capacity to cope – and even thrive, both of which hinge on an ability to dynamically adapt to life's changing situations. The forces that provide a positive (health-supporting) thrust are collectively known as Generalized Resistance Resources (GRRs). Along with GRRs, there are also Specific Resistance Resources (SRRs) – specific resources that reinforce specific human capabilities and are used in the face of stressors (Antonovsky, 1979). On the other side of the spectrum, the SOC will be hampered or weakened due to maladaptivity caused by inadequate resistance resources, or generalized resistance deficits (GRDs) (Golembiewski, 2017).

In the theory of salutogenesis, SRRs/GRRs are a finite resource, but GRDs are ubiquitous, emerging from constant changes of circumstance, knowledge, and our emotional attachments; from challenges; and from the irreversible entropy of life itself. The normative pathogenic approaches attempt to curtail this wave of stressors. As these stressors compound, they take a social, financial, and personal toll because controlling disease can be a fool's errand. The salutogenic model does not see the presence of a stressor as the cause of illness, but the lack of resources to deal with it. Salutogenic theorists do not advocate that diseases run unchecked, but that a robust SOC offers a bulwark should disease escape reasonable control measures.

A strong SOC means that a person will see the world as 'making sense', cognitively; as 'manageable', corporeally; and as 'meaningful', emotionally (Antonovsky, 1996). Inversely, a lack of meaning, understanding, or ability weaken the SOC. Thus, the SOC is where a person finds their sense of well-being once all the GRRs and SRRs minus all GRDs are tallied (Antonovsky, 1987). A GRD only matters when the GRRs and SRRs required to deal with it are absent or diminished. There is a dynamic relationship between SOC and resistance resources. People use resistance resources to make sense of the world, to meet demands, and to find meaning in these activities. Resistance resources strengthen the SOC (Antonovsky, 1987). The SOC is comprised of three intertwined dimensions – the cognitive (called 'comprehensibility'); the physical (called 'manageability'); and the affective (called 'meaningfulness') (Figure 22.1).

The dimensions of sense of coherence, 'comprehensibility', enable modelling of one's context in a way that agrees with reality (the unchanging rules of nature, such as the Newtonian laws), and with the normative perspectives of other people in the milieu, such as the laws and perspectives of one's society, family, or workplace.

'Manageability' is a person's ability to manage the corporeal domain. At its most basic, manageability is about functional existence – convenience, comfort, and somatic maintenance – the homeostasis of body temperature, blood glucose, hydration, levels of sleep, and even the regulation of neurotransmitters (Golembiewski, 2012). But manageability can be so much more: manageability enables one to manipulate and control their surroundings with expressive flair and imagination. If the individual feels that she is influencing what is happening around her, she does not perceive herself as a victim of circumstance (Antonovsky, 1991).

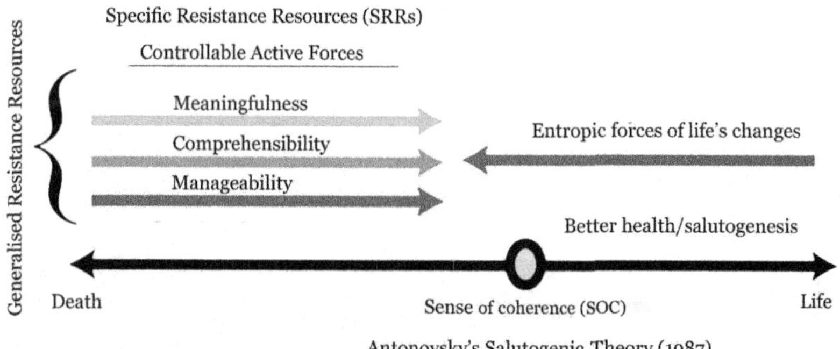

Against the ever-present reality of entropy and the ubiquitous demands of life, people struggle to maintain a sense of coherence (SOC); a relative feeling that all in all, things will work out for the best. The SOC is propelled forward by the resistance resources that support its three dimensions: manageability, comprehensibility, and meaningfulness.

FIGURE 22.1 Antonovsky's salutogenic theory.

'Meaningfulness' is possibly the most significant resource within a SOC. Antonovsky (1979) and Frankl (1963) describe meaning as the cornerstone of the need to live. It is what fills life with the impetus to endure. Meaningfulness is thus the motivational force of the SOC. It encourages one to combat entropy or even face death by transforming every challenge and all resistance against entropy into a worthwhile experience, so that adversity itself can become the fuel on which to thrive (Figure 22.2).

22.3 THE SALUTOGENIC METHOD

Salutogenesis is a model for understanding how social and environmental factors affect health. And wherever environmental factors affect health, the model helps discover why – whether designers consciously used a salutogenic paradigm to create the environment or not. But salutogenesis can be more than a post hoc tool for understanding environmental factors. Designers can use it purposefully. Wherever better health is a design objective, a thoughtful application of salutogenic principles is an excellent and practical approach to take, because it contains a hierarchical decision-making process to predict health effects, even when there is no empirical evidence available (Golembiewski, 2010). At its most simplified, the method calls architects to consider each design decision for the salutogenic effect it is likely to have, and for the inverse – for detrimental effects to health and well-being. This involves first assessing all design choices according to how much they contribute to the resources that enable resilience: comprehensibility, manageability, and meaningfulness.

A health-promoting built environment can focus on the SRRs that users may need in a specific milieu; people recovering from drug and alcohol addictions may need more meaningfulness in their lives, for example. It's important to note the intertwining nature of Resistance Resources though, and so resources

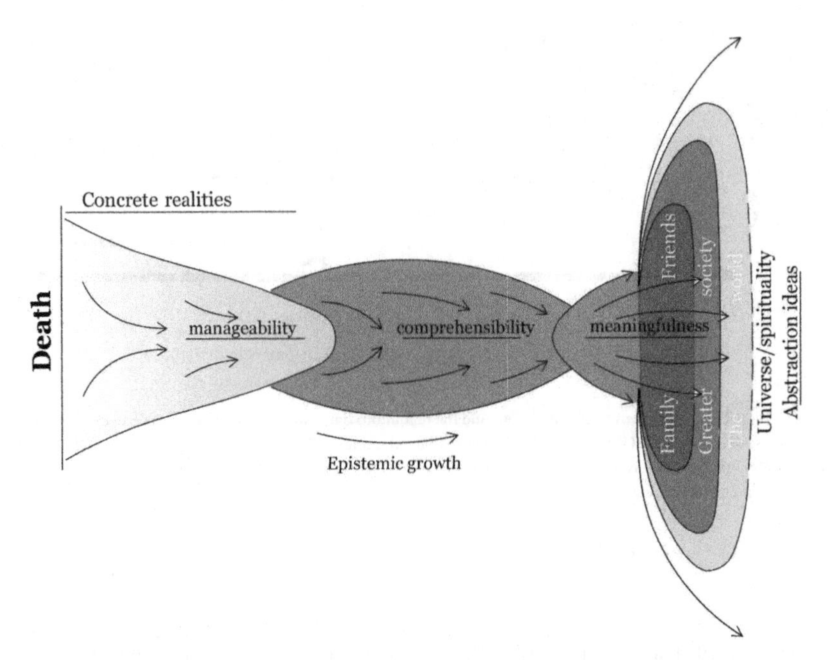

Each resistance resources occupies a different realm of relative corporeality. Manageability is solely concerned by concrete realities, whereas meaningfulness flourishes with greater abstractions and metaphysical ideas (which, being opposites to corporal realities, may not exist at all; in terms of psychology, it doesn't matter!) Comprehensibility works like a bridge to assist with achieving greater fulfilment, at first advancing manageability, then learning itself, and finally providing a scaffold of understanding to create meaningfulness.

FIGURE 22.2 The elements of sense of coherence in relation to one another.

that strongly support manageability and comprehensibility will still assist with meaningfulness deficits. For example, the structure of a building that protects us from weather becomes a GRR, because while it specifically assists with the manageability, it still helps provide the space needed to find meaning. When architects intentionally target a specific need of occupants with design, they target the SRRs.

22.4 ARCHITECTURE FOR MANAGEABILITY

Manageability is about finding the resources to live life itself; that is to meet both primary needs of survival such as having an appropriate shelter, being fed, staying clean, dry, feeling secure, and comfortable, access to medical care, etc. and the significant need of being in control of one's environment (Evans & Mccoy, 1998).

To strengthen manageability, at a minimum, designers must ensure that people's physical needs are met. Some physical parameters to ensure comfort and improve manageability include good air quality (ISO, 2005), access to natural daylight, as well as thermal, acoustic, and visual comfort (ISO, 2019). It does not sound like a high bar, but it is sometimes extraordinary just how designers ignore these basic needs!

There is ample evidence suggesting that design for manageability delivers a beneficial effect. Much of this is empirical, but with poorly understood causality: the effect of daylight on physiological and psychological well-being, for instance. Good sunlight is associated with positive health outcomes (Verderber, 1986; Lawson, 2001); it decreases the length of hospital stay for patients in psychiatric wards (Beauchemin & Hays, 1996); increases office workers' productivity, job satisfaction, and ultimately even reduces absenteeism (Leather et al., 1998). Similar interventions improve students' learning experience and school achievement (Schneider, 2002).[1]

Standards, guidelines, and quantitative assessments abound to ensure comfort for each architectural typology, and whilst these might be insufficient to reap full salutogenic benefits, they are without a doubt the most considered aspects of architectural design – not only by architects but also by their suppliers, academic researchers, and statutory and administrative bodies also. Considerations include ventilation; HVAC (Mathews et al., 2001); wall and window orientation, position, and measurement (Huizenga et al., 2006); materials that affect thermal, acoustic, and air quality; the acoustic properties of ceilings, walls, windows, and floors, and their claddings and coverings (Al Horr et al., 2016); and far too many other products and details to be listed.

There are several examples where design has gone beyond standards to address comfort and thereby improve manageability: Hiroshi Sambuich's Naoshima Hall in Naoshimaan (Japan) has an air tunnel as a variation of a traditional hip-and-gable roof to exploit natural ventilation. Baylor School of Business in Texas, USA, designed by Overland Partners Architects, permits natural light throughout the year and at the same time prevents adverse thermal heat gain, glare, and other issues by using meticulously designed skylight modules for the four-storey atrium. The modules have complementary orientations and angles to best respond to the sun's varying angles and provide the building with the optimal daylight (Figure 22.3).

Comfort is largely psychological and is not only about the *Goldilock's experience* (when the temperature, softness, etc. is 'just right') but also relates closely to people's ability to manipulate their environments. This is demonstrated by movable furniture layouts and spatial arrangements, which encourage social interaction (Sommer, 1969), and by a greater temperature tolerance if natural ventilation is directly alterable (Brager & de Dear, 1998). Privacy – the ability to regulate and manage social interactions – is of critical importance to a sense of control (Altman, 1975). To enable people to feel in control of their environments, residences must allow for opportunities to passively observe surroundings and to layer relative privacy from public areas through to the most intimate (Newman, 1996), and therefore create or inhibit social permeability according to residents' needs (Witte, 2003). It is interesting to note that

[1] Salutogenesis is not the only model that predicts these outcomes – we would also expect such outcomes from the Biophilia Hypothesis (Wilson, 1984) and the Attention Restoration Hypothesis (Kaplan, 1995), for instance. But whether salutogenesis is the only theoretical model that predicts health benefits is academic. It is the very nature of veracity that it can be arrived at from divergent perspectives. Importantly, Salutogenesis is broad enough to account for effects that the above theories cannot, such as comfort or religious affordance.

FIGURE 22.3 The atrium of Baylor University Paul L. Foster Campus for Business and Innovation designed by Overland Partners Architects.

Source: Photo by Mully Culver ©, Courtesy of Overland Partners Architects.

the deliberate denial of these opportunities is the most distinctive feature of the worst of prisons (Foucault, 1979) and hospitals (Serenko & Fan, 2013).

Architects can do a lot to improve comfort and convenience – but care must be taken that these do not incidentally erode higher-order salutogenic values. For example, smaller simpler houses are easy to 'manage', that is, to keep clean, navigate, and to heat and cool (Haugan et al., 2015) but they might not be suitable for appropriate symbolic representation (see 'comprehensibility'), for keeping pets, entertaining other people, or indulging one's passions (see 'meaningfulness') or other undefined needs. For this reason, we can only predict general positive outcomes for smaller dwellings with mixed results. But even with this caveat, the kinds of things an architect can do to improve manageability are endless. From the smaller spaces concept (above) to innovations like double dishwasher kitchens (where users keep one for storage, and the other for cleaning. This system removes the need to empty the dishwasher into cupboards, thus saving time and effort). Other issues of convenience are related to the scale of the users. Psychological Design is designing a primary school with little (but functional) kitchens in each classroom scaled for the children who will be using them. Gender can also become an issue of manageability because facilities like cupboards are often scaled for men (Ainley, 1998).

22.5 ARCHITECTURE FOR COMPREHENSIBILITY

Comprehensibility in architecture is the ability to make sense and negotiate the physical environment. Comprehensible architecture translates material existence into an empowering experience, which nurtures our sense of individuality, identity, and sense of efficacy.

At a minimum, a designed environment must facilitate wayfinding, project the right messages for the user (in terms of typology and semiotics), and provide self-explanatory and simple controls, equipment, and customisation features. But comprehensibility in buildings can be so much more. To an extent, comprehensibility depends on the building's typology and purpose. A question must first be asked: 'What do users want to do in this building, and what extra steps can the designer take to assist them?'.

When thinking about wayfinding, designers must consider signage as a partial solution because it is problematic (it is sometimes inaccurate, language-specific, subject to interpretation and sometimes conflicts with other cues such as known landmarks, verbal advice, and is agnostic about other orientation cues such as time of day). Furthermore, signage carries a cognitive load, thereby adding to unnecessary GRDs that salutogenic designers are well-advised to avoid. People are guided by distant landmarks and distinctive features within the space (Lynch, 1960), so, windows that provide visual connections to the weather, time of day, and external distant landmarks, such as identifiable buildings or other prominent features, support holistic orientation. Good, logical design also helps, especially in institutional scale buildings – but also in houses. Landmarks can also be of assistance inside, especially for visitors and people with memory impairments. A configuration of multiple cues, both distant and nearby (such as unique architectural design features or stationary decorations), help orientation a great deal (Sternberg & Wilson, 2006). ZGF Architects, the designers of the Seattle Children's Hospital, and Medical Architecture, the architects of Ferndene Children and Young Peoples Inpatient Service chose to tell a story through the illustrative artwork that reinforces identification for each zone. This helps with navigation (Figure 22.4).

Authenticity in design assists people with comprehensibility because it is the very opposite of ambiguousness. It confirms rather than challenges people's perceptions, thereby allowing people to focus on their own needs rather than the trickery of the building they are in (Golembiewski, 2010). Places with an authentic atmosphere inspire people and draws them into some kind of relationship (Friedman, 2010), thereby also reaching into the sphere of meaningfulness. To form this relationship,

The animals and colours are a part of a carefully crafted navigation system that helps children and teenagers find themselves in a story and also add to the sense of play and joy.

FIGURE 22.4 Ferndene Children and Young Peoples Inpatient Service, designed by Medical Architecture.

Source: Photo courtesy of Jill Tate ©.

it is important to note that prior experiences and associations are part of how people come to understand spaces (Albright, 2015). We perceive the environment through memory, by reflecting on familiar concepts (knowledge); through one's state of being (ontology); through language, culture, the affordances that spaces and objects offer, and through known forms, materials, textures, emotions, expectations, and fundamental perceptual ability (Golembiewski, 2010). Understanding this, the

The project breaks conventions by using a lot of wood in the interior of the hospital along with indigenous art to express the indigenous values and associations of the region. The hospital also has large glass walls and open corridors to let in lots of light, thereby deinstitutionalizing the environment.

FIGURE 22.5 The St Mary's Sechelt Hospital designed by Farrow Partners Architects and Perkins+Will.

Source: Photo by Andrew Latreille ©, courtesy of Farrow Partners Architects.

proactive use of memory, culture, and language reminds people that a place is built especially for them and in honour of their values. Farrow Partnership in association with Perkins+Will considered this issue in designing the Sechelt Hospital. The extensive use of wood (which is essential to the Indigenous culture of the region) and Indigenous art bring local Indigenous culture and tradition into the building. Patients, staff, and members of the community instantly recognise these artefacts: they contribute hugely to the creation of a sense of place in a typology (hospitals) that all too easily become generic (Figure 22.5).

Again, comprehensibility is empowering. This empowerment is dependent on the designed environment that structures and suggests narratives, and these reflect and reinforce the personal narratives of occupants – for example, healthcare architecture can fulfil a promise of growth and restoration and replace the normative institutional language of illness and treatment. In Sechelt Hospital, for instance, Farrow Partners in association with Perkins+Will architects followed contemporary trends to deinstitutionalise by designing a hospital filled with natural daylight and good views throughout single patient bedrooms and operable windows. Together these say, 'this place is about you and your recovery'. Similarly, in the Al Wakra Centre for Respite and Recovery, a psychiatric facility planned for Qatar (designed by MAAP, MA, and AECOM), wonderful opportunities exist for patients to forget that they are in an institution. Structured around gardens of every scale from personal ones attached to bedrooms to bigger ones attached to pods, and bigger ones yet where psychiatric patients can train horses and birds, go for long walks in enchanting gardens, and eat in 'restaurants' rather than mess-halls or their bedrooms (Figure 22.6).

Another example of a strong congruence between message and architectural design is the Living Planet Centre (the UK headquarters of the World Wildlife Fund – WWF) designed by Hopkins architects. This sustainable, green building establishes a convincing narrative that matches the brand of WWF (Figure 22.7). The high ceilings and

A wide range of opportunities to access gardens, to enjoy interacting with nature and animals, and the resort-like environment remove institutional feeling in design and promote well-being and recovery.

FIGURE 22.6 Al Wakra Centre for Psychiatric Respite and Rehabilitation designed by MAAP, MA, and AECOM.

Source: Image courtesy of Jan A. Golembiewski (Knowledge Lead and co-designer).

The building is designed to constantly remind employees and visitors of the WWF mission, which is to build a future in which humans live in harmony with nature and ensuring that the use of renewable natural resources is sustainable.

FIGURE 22.7 WWF-UK's Living Planet Centre by Hopkins Architects.

Source: Photo by Airey Spaces ©, courtesy of Airey Spaces.

open-atrium layout designed around three huge weeping figs connect the occupants with the natural environment, which is at the core of WWF's mission. Similarly, the outdoor space allows employees access to flowers and produce, and this speaks to the company's sustainable values.

Comprehensibility is closely related to perception itself, a fundamental part of the early cognitive processes. The Ecological Perception Theory (Gibson, 1979) is an ideal model for designers to understand perception because this model recognises that perception is not passive but involves both the whole multi-sensory organism and the environmental context, from which it cannot be meaningfully separated (Gibson, 1979; Mallgrave, 2014). The primary mode of perception is not vision, hearing, or a sense of touch (a normative belief) but *opportunities to act – to do or say or think something*. Designers can structure these opportunities (called affordances) into the built environment. To foster comprehensibility, we must understand affordances and how the body is always in communion with environmental context – for most of us, most of the time, this means that we must better understand the built environment (Golembiewski, 2016).

22.6 ARCHITECTURE FOR MEANINGFULNESS

With a rich sense of meaningfulness, one can encounter dreadful circumstances – pain, illness, insufficiency of resources, etc., and feel confident that, eventually, everything will turn out for the best (Frankl, 1963; Antonovsky, 1987).

People develop their sense of meaning spontaneously. And against this backdrop, our built environment tends to do more harm than good. So, before we look at how designers use the built environment to deterministically strengthen meaningfulness for users, we should look at how meaningfulness erodes. Our cities, hospitals, and other public buildings and public housing are replete with *meaninglessness*. There are mean-spirited and thoughtless acts everywhere. Locked doors that should be open, aggressive, and polarising symbols and signs, dead-ends, harsh lighting, nasty odours, and visual and acoustic noise – or even filth. There are empty spaces, sidewalks that stop before multi-lane highways, barbed wire fences, CCTV cameras, and huge open-air carparks that create voids between buildings through which people may pass through. In short, all design where the consideration of users of the space (either insiders or visitors) is secondary (or disregarded altogether) to the primary functions of the space. System-cantered and otherwise careless design can transform to person-cantered alternatives, lest it destroy our innate sense of meaningfulness.

Once architectural designers identify meaninglessness and remove it, they can enhance meaningfulness using a couple of basic tools: the first is *quality*. The provision of quality surroundings tells people that they are worthy and that they are in a good environment. The second (and no less important) is the provision of *wholesome choices* of things to do or get engaged with. Choices make us feel good because a good choice is the outcome of an evolutionary highpoint. Throughout our evolutionary history, we have had to struggle to do what needs to be done: to protect our families and flocks, to bring in the grain, to pay our bills. But when the grain is siloed, there is suddenly nothing pressing to do, and it is time to celebrate and acknowledge our prosperity. We cannot underestimate the motivational power of meaningful choices. Whereas manageability and comprehensibility enable action, meaningfulness drives that action, and if we – as people – are to achieve anything (including better health), we must do so actively (Smithies & Webster, 2018), because, in salutogenic terms, the GRDs (as previously mentioned) are ubiquitous and entropic.

Beyond these two maxims (quality and choice), we can also look at how people find meaning in their lives and support them specifically. Meaning is found in all the things we care most about – the things we 'live and die for'. It is understood, therefore, that other people are a mainstay for meaningfulness for most everyone and we can use sociopetal design to support social interactions (Shepley & Pasha, 2013). Animals are also important, so we can design special provisions for animals. People find meaning in religion and other metaphysical abstractions, so we can design provisions for religious and spiritual activities (Snodgrass, 1990). People find meaning in self-expression, so we design support for art music, poetry, etc. (Dilani, 2005). Affordances for these activities are becoming common in Dementia Care facilities (Fleming et al., 2020; Zeisel, 2020). And finally, people find meaning, peace, and solace in nature, and that means integration with the natural environment (this is the main premise of both the biophilia hypothesis of Wilson [1984] and the Attention Restoration Theory of Kaplan [1995]).

When designing, it is important to think beyond the primary target users and consider their visitors and friends also. So, when designing for aged care, think also of children's playgrounds, communal activities, and simply the opportunity to see and hear other people. It is in these interactions that the needs for contact, knowledge, and stimulation are satisfied, which is truly meaningful and important to humanity (Gehl, 1987). As an example, Levy Park in Houston (Texas, USA) offers free programming and public events. The park offers both vast and intimate scaled spaces to provide something for everyone. It is the vitality and diversity of the park that attracts people of all ages and backgrounds, and undoubtedly, this vitality is a product of the abundance of wholesome affordances (Ziegler, 2014).

Nature offers positive aesthetic affordances on all levels: people can feel themselves react to an assortment of colours, forms, scents, and sounds, and this encourages humans to let go of other worries and concerns and 'reboot' the attentional system (Kaplan & Kaplan, 1989; Kaplan, 1995; Herzog et al., 2003). The integration of nature and built environment is also a significant source of meaningful experience for users, and therefore a considerable effect on their sense of coherence. In this regard, biophilic design can be seen not as a competing theory, but as an important strategy of salutogenesis in promoting well-being in built spaces (Mazuch, 2017; Abdelaal & Soebarto, 2019). The influence of nature on physical and mental well-being, including recovery from illness and improved productivity at work, is among the most studied design for health interventions (e.g., Ulrich, 1984). As such, well-designed buildings now routinely seek to promote the connection with nature. Maggie's Cancer Centre in Manchester (New York, USA), designed by Foster + Partners, provided a greenhouse for patients to gather, work with their hands, and enjoy the healing power of nature, light, and the outdoors. This gives the patients a sense of meaningfulness at a time when they may feel at their most vulnerable (Figure 22.8).

The integration of nature into urban settings means that pedestrians and nearby buildings alike will benefit from enhanced meaningfulness. Many neglected and undefined areas in cities that undergo landscape redesign have witnessed notable

The greenhouse of the center has created a great environment to contemplate, get together, and cultivate plants. It also provides aesthetic pleasure.

FIGURE 22.8 The greenhouse of Maggie's Cancer Centre in Manchester designed by Foster + Partners.

Source: Garden designed by Dan Pearson. Photo by Philip Durrant ©, courtesy of Philip Durrant.

changes, including the better and more meaningful social interaction among city residents. As an example, the Tianjin Bridged Gardens in Tianjin City (China) turned a heavily polluted and deserted area into an environment that celebrates the local culture, landscapes, and provides recreational opportunities for the surrounding communities.

Interaction with animals has therapeutic value, especially for children, elderly, and people tackling the mental and physical health problems because animals are good proxies for other people and are often more reliable! Interaction with animals also improves moods and even indicators of their somatic health. Understanding this, pet therapy is used in residential and health canters in the process of rehabilitation (Bivens et al., 2007; Creagan et al., 2015). Designing for animal companionship might be practical – like designing interior finishes to make cleaning animal mess easy or it might mean special provisions in a hospital to allow animal visitors such as places where animals can visit safely, where the risks of ultrasonic noises and other shocking stimuli are low. Such measures have been incorporated into many healthcare centres, including the Mayo Clinic, in Rochester New York (Creagan et al., 2015). Chase Memorial Nursing Home in New York has also considered the adequate provision for animals – cats, dogs, birds in the centre. This approach in design and treatment has resulted in improvements in the daily performance of the elderly (Gawande, 2014). The proposed Al Wakra Center for Psychiatric Respite and Rehabilitation has horse stables, horse lunging arenas, and walking trails, as well as an aviary for keeping pet birds. VAC library in Hà Nội of Vietnam is another interesting example of combining meaningful elements in design (Figure 22.9). The project is an open climbable wooden structure containing a small chicken coop. There is also a garden and a fishpond. Being in the urban area, the designers aimed to carry the traditional Vietnamese horticulture, aquaculture, and animal husbandry, from rural to urban areas. So, the project is a part library, part urban farm, part playground.

FIGURE 22.9 VAC Library in Dương Nội, Hà Đông, Hà Nội, Vietnam, by Farming Architects.

Source: Photo by Nguyễn Thái Thạch; An Việt Dũng©, courtesy of image by Farming Architects.

It might be harder to design for meaningfulness than for the other salutogenic resources, but as we can see in these examples, the effort is worth it: it invariably turns a project into an exemplar that is worth imitating. As difficult as it might be to design for meaningfulness, it is also the most open-ended and creative sphere of design. The constructed environment can allow users to feel in control over their ecological footprints, it can encourage food production, and so much more. Precht Architects step beyond what others in industry considered normal to develop a concept for modular housing (the conceptual Farmhouse Project), where residents will be able to produce their food in vertical farms (Figure 22.10). The plan is to enable residents of a city (Toronto) to have control over what they consume, and that will be very fortifying for those who choose to live in the units.

Participatory processes of design and construction can also add to a sense of meaning. Case studies of participatory architecture show that engaging the end users in the process of design and construction leads to numerous positive consequences such as mutual learning (learning one another's knowledge and skills), cultural exchange (among participants), developing a sense of belonging to the community, and sense of connection to the product (Di Mascio & Dalton, 2017; Lahtinen et al., 2020). More importantly, the co-creation of something tangible (either a house or a public building like a community hall) allows the users to feel worthwhile as they make decisions

The project will integrate nature, choices, quality, and people's living spaces to bring meaningfulness back into urban life.

FIGURE 22.10 The Farmhouse Project concept proposed by Precht Architects.

Source: Image courtesy of Studio Precht: www.precht.at

and shape their future. The sum of all these feelings is central to meaningfulness, and this makes participatory architecture an important method of salutogenic design.

22.7 CONCLUSION

The normative view on somatic health recognises physical influences as key drivers, such as pathogens, diet, and exercise, and genetics and the normative thought on mental health issues causation from the mind – so people will naturally be sceptical about how such things as architecture and urban design affect these diverse aetiologies. Salutogenesis will never be the magic bullet presented in marketing literature. It cannot solve all the complexity of mental and somatic illness, infection control, poverty, or environmental degradation. But by understanding salutogenic principles, architects can make reasonable predictions and assertions about how they engage with the complexity at the core of these issues. Salutogenic architecture is not medicine or an environmental panacea – it is the physical embodiment of caring, and, as such, against medicine in the same way that 'nursing' is – that is, as essential support. Antonovsky, the founder of salutogenic theory, never intended to displace medicine and other paradigms of thought. Instead, the theory arose from a need to understand the natural tendency towards wellness and the natural process of recovery.

Very few of the projects mentioned above considered salutogenesis explicitly (the projects by Farrow and Partners, MAAP, and Psychological Design being notable exceptions), instead the architects designed most projects using allied values: human-centredness, a passion for quality, and an interest in the well-being of users. The projects mentioned here are not an exhaustive list, nor are their inventions. These are projects in which salutogenic features are recognisable and we chose these to illustrate how salutogenesis works, with or without knowledge of the theory. You do not need to know the chemical and thermal dynamics of rapid oxidisation of fuels to start a fire (although in terms of chemistry, that is indeed what a fire is), but knowledge makes it easier to identify where problems arise (why won't the fire light when it is underwater?) and to identify new fuel sources (we cannot burn water, but perhaps we can burn other liquids?) A close look at these projects suggests that their design characteristics reflect the three salutogenic tools: meaningfulness, comprehensibility, and manageability, mostly without even knowing the theory. The fact that those who have never even heard of salutogenesis can identify successful projects only highlights the validity of the theory and salutogenic approaches. But now, armed with the theory of salutogenic theory, a new bar can be set, for better, even more human-cantered, even healthier buildings to follow.

22.8 SUGGESTIONS FOR FUTURE SALUTOGENIC RESEARCH AND PRACTICE

The main reason that architects do not consider salutogenesis in their design is that it is virtually unheard of except in very niche areas of practice. Educating health professionals about salutogenesis has meaningfully contributed to the practice of salutogenesis in the health sector (Vinje et al., 2017). To deliver healthier architecture, we should also consider teaching salutogenic theory in universities and encourage architectural students to reflect on salutogenic and person-centred approaches to design. Advocates should promote salutogenesis to raise awareness among policy-makers, stakeholders, users, and others vested in architectural projects.

Determining what contributes to meaningfulness, comprehensibility, and manageability may be challenging: these three dimensions cover a broad range of overlapping interventions (as shown in Figure 22.2) and categorising them is an act of interpretation (Mazzi, 2021). But the task is simply to think of users in the designed context. There is no single list of comprehensible, manageable, and meaningful interventions for all typologies, but the resources that salutogenic architecture calls on is a language, and like all languages, they address diverse groups of people differently. For instance, can a space intended to fortify meaningfulness work equally given different religions, religious symbology, and taboos? Antonovsky emphasises that meaningfulness is culturally encoded, and substantive answers to how it can be fostered will vary from culture to culture and from situation to situation (Antonovsky, 1996). It helps to know about the users. Architects can scrutinise the concerns that users will bring to the picture means developing an awareness of the users' socio-economic and cultural backgrounds, feelings, expectations, and preconceptions. This sounds obvious, but it is a U-turn from the mode of the twentieth century, which sought universal solutions to diverse cultures. This insight will enable architects to ask the right questions for the right population, and therefore implement the right interventions for those users (Ziegler, 2014). This requires involving the entire system and as many stakeholders as possible (ideally with some basic knowledge of the salutogenic theory) so that everyone involved understands what they are contributing and what compromises are being made and why.

To expand our knowledge of the salutogenic impact of architectural interventions, we should continue to research the dynamic relationship between the end users' sense of coherence and the affordances of each architectural typology, and of the discipline. To achieve this, we can continue as we have here – by interpreting successful exemplars and existing evidence-based research from a salutogenic perspective. But we need not stop there. We should also look at reframing successful interventions from unique typologies and disciplines (Dorst, 2015). This can be done with a specific lens to identify salutogenic elements, just as Chrysikou et al. (2018) have done when applying hotel design approaches to dementia care.

22.9 IMPLICATIONS FOR PRACTICE

- Designers can use **salutogenesis** to quickly scrutinise decisions and to make assertions about how they can best improve health and well-being in architectural environments.
- **Manageability** improves in this model by enhancing basic human needs such as comfort, convenience, safety, and privacy.
- Designers can enhance **comprehensibility** among users by making environments more understandable, coherent, and authentic. Supporting people's desired narratives also improves environments.
- **Meaningfulness** is the most important salutogenic resource. Designers support this by identifying and removing anything that speaks of its

opposite – meaninglessness (things such as dead ends, windowless rooms, unhelpful signage, and design-to-rule decisions like minimum ceiling heights, room sizes, and cheap, careless design choices).

- **Person-centred design** props meaningfulness; that is, design that embodies 'care'. People find meaningfulness in all design features that make life worth living, including nudges towards wholesome choices; opportunities for self-expression; growth; social connectivity; connection with animals; and integration with nature.

REFERENCES

Abdelaal, M. S., & Soebarto, V. (2019). Biophilia and salutogenesis as restorative design approaches in healthcare architecture. *Architectural Science Review*, *62*(3), 195–205. https://doi.org/10.1080/00038628.2019.1604313

Ainley, R. (1998). *New frontiers of space, bodies, and gender*. Routledge.

Albright, T. D. (2015). Neuroscience for architecture. In S. Robinson & J. Pallasmaa (Eds.), *Mind in architecture: Neuroscience, embodiment, and the future of design* (pp. 197–219). The MIT Press. http://ebookcentral.proquest.com/lib/qut/detail.action?docID=3339978

Al Horr, Y., Arif, M., Katafygiotou, M., Mazroei, A., Kaushik, A., & Elsarrag, E. (2016). Impact of indoor environmental quality on occupant well-being and comfort: A review of the literature. *International Journal of Sustainable Built Environment*, *5*(1), 1–11. https://doi.org/10.1016/j.ijsbe.2016.03.006

Altman, I. (1975). *The Environment and social behavior: Privacy, personal space, territory, and crowding*. Brooks-Cole.

Antonovsky, A. (1979). *Health, stress, and coping*. Jossey-Bass.

Antonovsky, A. (1987). *Unravelling the mystery of health: How people manage stress and stay well*. Jossey-Bass.

Antonovsky, A. (1991). The structural sources of salutogenic strengths. In C. L. Cooper & R. Payne (Eds.), *Personality and stress: Individual differences in the stress process* (pp. 67–104). John Wiley & Sons.

Antonovsky, A. (1996). The salutogenic model as a theory to guide health promotion. *Health Promotion International*, *11*(1), 11–18. https://doi.org/10.1093/heapro/11.1.11

Beauchemin, K. M., & Hays, P. (1996). Sunny hospital rooms expedite recovery from severe and refractory depressions. *Journal of Affective Disorders*, *40*(1), 49–51. https://doi.org/10.1016/0165-0327(96)00040-7

Bivens, A., Leinart, D., Klontz, B., & Klontz, T. (2007). The effectiveness of equine-assisted experiential therapy: Results of an open clinical trial. *Society & Animals*, *15*(3), 257–267. https://doi.org/10.1163/156853007X217195

Brager, G. S., & de Dear, R. J. (1998). Thermal adaptation in the built environment: A literature review. *Energy and Buildings*, *27*(1), 83–96. https://doi.org/10.1016/S0378-7788(97)00053-4

Brick, S. (2020). *Salutogenic design*. Retrieved August 1, 2020, from www.stephaniebrick.com/salutogenic-design

Chrysikou, E., Tziraki, C., & Buhalis, D. (2018). Architectural hybrids for living across the lifespan: Lessons from dementia. *The Service Industries Journal*, *38*(1–2), 4–26. https://doi.org/10.1080/02642069.2017.1365138

Creagan, E. T., Bauer, B. A., Thomley, B. S., & Borg, J. M. (2015). Animal-assisted therapy at Mayo Clinic: The time is now. *Complementary Therapies in Clinical Practice*, *21*(2), 101–104. https://doi.org/10.1016/j.ctcp.2015.03.002

Dilani, A. (2001). Psychosocially supportive design – Scandinavian healthcare design. In A. Dilani (Ed.), *Design and health – the therapeutic benefits of design* (pp. 31–38). AB Svensk Byggtjänst.

Dilani, A. (2005). A new paradigm of design and health in hospital planning. *World Hospitals and Health Services: The Official Journal of the International Hospital Federation, 41*(4), 17–21.

Dilani, A., & Armstrong, K. (2008). The salutogenic approach – designing a health-promoting hospital environment. *World Hospitals and Health Services, 44*(3), 32–35.

Di Mascio, D., & Dalton, R. (2017). Using serious games to establish a dialogue between designers and citizens in participatory design. In M. Ma & A. Oikonomou (Eds.), *Serious games and edutainment applications*. Springer. https://doi.org/10.1007/978-3-319-51645-5_20

Dorst, K. (2015). Frame creation and design in the expanded field. *She Ji: The Journal of Design, Economics, and Innovation, 1*(1), 22–33. https://doi.org/10.1016/j.sheji.2015.07.003

Evans, G. W., & McCoy, J. M. (1998). When the buildings don't work: The role of architecture in human health. *Journal of Environmental Psychology, 18*(1), 85–94. https://doi.org/10.1006/jevp.1998.0089

Farrow, T. (2019). *Causing health: A design talk with Tye Farrow*. https://www.mca.ie/news/causing-health-a-design-talk-with-tye-farrow/

Fleming, R., Zeisel, J., & Bennett, K. (2020). *World Alzheimer report 2020: Design, dignity, dementia: Dementia-related design and the built environment* (Vol. 2). www.alzint.org/resource/world-alzheimer-report-2020/

Foucault, M. (1979). *Discipline and punish: The birth of the prison*. (Trans A. Sheridan). Vintage.

Frankl, V. E. (1963). *Man's search for meaning: An introduction to logotherapy*. Pocket Books.

Friedman, A. (2010). *A place in mind: The search for authenticity*. Véhicule Press.

Gawande, A. (2014). *Being mortal: Illness, medicine, and what matters in the end*. Profile Books.

Gehl, J. (1987). *Life between buildings: Using public space*. Van Nostrand Reinhold.

Gibson, J. J. (1979). *The ecological approach to visual perception*. Houghton Mifflin Company.

Golembiewski, J. A. (2010). Start making sense: Applying a salutogenic model to architectural design for psychiatric care. *Facilities (Bradford, West Yorkshire, England), 28*(3–4), 100–117. https://doi.org/10.1108/02632771011023096

Golembiewski, J. A. (2012). Salutogenic design: The neurological basis of health-promoting environments. *World Health Design: Architecture, Culture, Technology, 5*(3), 62–69.

Golembiewski, J. A. (2016). The designed environment and how it affects brain morphology and mental health. *HERD, 9*(2), 161–171. https://doi.org/10.1177/1937586715609562

Golembiewski, J. A. (2017a). Salutogenic architecture in healthcare settings. In M. B. Mittelmark, S. Sagy, M. Eriksson, G. F. Bauer, J. M. Pelikan, B. Lindstrom, & G. A. Espnes (Eds.), *The handbook of salutogenesis* (pp. 267–275). Springer/Nature Group.

Golembiewski, J. A. (2017b). Salutogenics and residential care for people with dementia. *Australian Journal of Dementia Care, 6*(3), 25–28.

Halpern, D. (1995). *Mental health and the built environment: More than bricks and mortar?* Taylor & Francis.

Haugan, G., Woods, R., Høyland, K., & Kirkevold, Ø. (2015). *Er smått alltid godt i eldreomsorgen? Kunnskapsstatus om botilbud*. SINTEF Akademiske Forlag. ISBN 978-82-536-1447-2(pdf)

Hedge, A. (2017). *Ergonomic workplace design for health, wellness, and productivity*. CRC Press.

Herzog, T. R., Colleen, Maguire, P., & Nebel, M. B. (2003). Assessing the restorative components of environments. *Journal of Environmental Psychology, 23*(2), 159–170. https://doi.org/10.1016/S0272-4944(02)00113-5

Huizenga, C., Zhang, H., Mattelaer, P., Yu, T., Arens, E. A., & Lyons, P. (2006). *Window performance for human thermal comfort*. Center for the Built Environment. https://escholarship.org/uc/item/6rp85170

International Organization for Standardization. (2005). *Ergonomics of the thermal environment*. (ISO Standard No. 7730:2005). ISO.

International Organization for Standardization. (2019). *Building environment design – indoor environment – daylight opening design for sustainability principles in the visual environment* (ISO Standard No. 19454:2019). ISO.

Kaplan, S. (1995). The restorative benefits of nature: Toward an integrative framework. *Journal of Environmental Psychology*, *15*(3), 169–182. https://doi.org/10.1016/0272-4944(95)90001-2

Kaplan, S., & Kaplan, R. (1982). *Cognition and environment: Functioning in an uncertain world*. Praeger.

Kaplan, R., & Kaplan, S. (1989). *The experience of nature: A psychological perspective*. Cambridge University Press.

Lahtinen, M., Sirola, P., Peltokorpi, A., Aalto, L., Kyrö, R., Salonen, H., Ruohomäki, V., & Reijula, K. (2020). Possibilities for user-centric and participatory design in modular health care facilities. *Intelligent Buildings International (London)*, *12*(2), 100–114. https://doi.org/10.1080/17508975.2018.1512470

Lawson, B. (2001). *The language of space*. Architectural Press.

Leather, P., Pyrgas, M., Beale, D., & Lawrence, C. (1998). Windows in the workplace: Sunlight, view, and occupational stress. *Environment and Behavior*, *30*(6), 739–762. https://doi.org/10.1177/001391659803000601

Lynch, K. (1960). *The image of the city*. MIT Press.

Lyon, C. (2017). Humanist principles, sustainable design and salutogenics: A new form of healthcare architecture. *Architectural Design*, *87*(2), 56–65. https://doi.org/10.1002/ad.2153

Maass, R., Lillefjell, M., & Espnes, G. A. (2017). The application of salutogenesis in cities and towns. In M. B. Mittelmark, S. Sagy, M. Eriksson, G. F. Bauer, J. M. Pelikan, B. Lindstrom, & G. A. Espnes (Eds.), *The handbook of salutogenesis* (pp. 171–179). Springer/Nature Group. https://doi.org/10.1007/978-3-319-04600-6_18

Mallgrave, H. F. (2014). Cognition in the flesh . . . the human in design. *Thresholds*, *42*, 76–87.

Mathews, E. H., Botha, C. P., Arndt, D. C., & Malan, A. (2001). HVAC control strategies to enhance comfort and minimise energy usage. *Energy and Buildings*, *33*(8), 853–863. https://doi.org/10.1016/S0378-7788(01)00075-5

Mazuch, R. (2017). Salutogenic and biophilic design as therapeutic approaches to sustainable architecture. *Architectural Design*, *87*(2), 42–47. https://doi.org/10.1002/ad.2151

Mazzi, A. (2021). Toward a unified language (and application) of salutogenic design: An opinion paper. *HERD*, *14*(2), 337–349. https://doi.org/10.1177/1937586720967347

Newman, O. (1996). *Creating defensible space*. US Department of Housing and Urban Development, Office of Policy Development and Research.

Olakitan Atanda, J. (2019). Developing a social sustainability assessment framework. *Sustainable Cities and Society*, *44*, 237–252. https://doi.org/10.1016/j.scs.2018.09.023

Pentecost, R. (2013). *A perfect storm: The economic challenge and the salutogenic response*. International Academy for Design and Health, Keynote Address, 9th Design & Health World Congress & Exhibition, Brisbane.

Roskams, M., & Haynes, B. (2020). Salutogenic workplace design: A conceptual framework for supporting sense of coherence through environmental resources. *Journal of Corporate Real Estate*, *22*(2), 139–153. https://doi.org/10.1108/JCRE-01-2019-0001

Ruohomäki, V., Lahtinen, M., & Reijula, K. (2015). Salutogenic and user-centred approach for workplace design. *Intelligent Buildings International (London)*, *7*(4), 184–197. https://doi.org/10.1080/17508975.2015.1007911

Schneider, M. (2002). *Do school facilities affect academic outcomes?* National Clearinghouse for Educational Facilities.

Serenko, N., & Fan, L. (2013). Patients' perceptions of privacy and their outcomes in health-care. *International Journal of Behavioural and Healthcare Research, 4*(2), 101–122. https://doi.org/10.1504/IJBHR.2013.057359

Shepley, M. M., Pasha, S., Ferguson, P., Huffcut, J. C., Kiyokawa, G., & Martere, J. (2013). *Design research and behavioral health facilities* (pp. 1–81). The Center for Health Design.

Smithies, J., & Webster, G. (2018). *Community involvement in health: From passive recipients to active participants*. Routledge.

Snodgrass, A. (1990). *Architecture, time, and eternity: Studies in the stellar and temporal symbolism of traditional buildings* (Vol. 1). Aditya Prakashan.

Sommer, R. (1969). *Personal space: The behavioral basis of design*. Prentice-Hall.

Sternberg, E. M., & Wilson, M. A. (2006). Neuroscience and architecture: Seeking common ground. *Cell, 127*(2), 239–242. https://doi.org/10.1016/j.cell.2006.10.012

Ulrich, R. S. (1979). Visual landscapes and psychological well-being. *Landscape Research, 4*(1), 17–23. https://doi.org/10.1080/01426397908705892

Ulrich, R. S. (1984). View through a window may influence recovery from surgery. *Science (American Association for the Advancement of Science), 224*(4647), 420–421. https://doi.org/10.1126/science.6143402

Vaandrager, L., & Kennedy, L. (2017). The application of salutogenesis in communities and neighborhoods. In M. B. Mittelmark, S. Sagy, M. Eriksson, G. F. Bauer, J. M. Pelikan, B. Lindstrom, & G. A. Espnes (Eds.), *The handbook of salutogenesis* (pp. 159–170). Springer/Nature Group. https://doi.org/10.1007/978-3-319-04600-6_17

Verderber, S. (1986). Dimensions of person-window transactions in the hospital environment. *Environment and Behaviour, 18*(4), 450–466. https://doi.org/10.1177/0013916586184002

Vinje, H. F., Ausland, L. H., & Langeland, E. (2017). The application of salutogenesis in the training of health professionals. In M. B. Mittelmark, S. Sagy, M. Eriksson, G. F. Bauer, J. M. Pelikan, B. Lindstrom, & G. A. Espnes (Eds.), *The handbook of salutogenesis* (pp. 267–275). Springer/Nature Group. https://doi.org/10.1007/978-3-319-04600-6_29

Wilson, E. O. (1984). *Biophilia: The human bond with other species*. Harvard University Press.

Witte, N. (2003). *Privacy: Architecture in support of privacy regulation* [M.Sc. thesis, University of Cincinnati].

Xie, H., Clements-Croome, D., & Wang, Q. (2017). Move beyond green building: A focus on healthy, comfortable, sustainable, and aesthetical architecture. *Intelligent Buildings International (London), 9*(2), 88–96. https://doi.org/10.1080/17508975.2016.1139536

Zeisel, J. 2020. *At home with growing older presents John Zeisel, 'Despair and Hope': Innovations in the design of care environments and care approaches for people living with dementia*. Retrieved, from www.athomewithgrowingold.com.

Ziegler, E. (2014). *Application of a salutogenic design model to the architecture of low-income housing (T)*. University of British Columbia. Retrieved, from https://open.library.ubc.ca/collections/ubctheses/24/items/1.0135612

23 Author Reflections

Jan A. Golembiewski and Narges Farahnak Majd

23.1 PERSONAL REFLECTIONS BY NARGES FARAHNAK MAJD

I learned of the salutogenesis theory around nine years ago while working on my master thesis through my supervisor's research work – Dr Jan A. Golembiewski. The theory inspired my research immensely: it was easy to grasp, all-inclusive, and could effectively contribute to a health-promoting environment. For my thesis project, I studied the influence of the built environment on mental health patients, for which I spent a year and a half on and off in a mental health facility in Iran, observing the patients' routines (their interaction with one another and their environment) and interviewing staff. This observation period suggested that the physical and social resources fostering the sense of coherence – the core concept in salutogenic theory comprising three dimensions: meaningfulness, comprehensibility, and manageability – can impact patients' sense of well-being.

Being fascinated by the salutogenic model and its implications in design, I started raising it with my classmates and professors. I also once talked about the theory with the occupational therapist of the facility I used to visit. He listened passionately and commented: 'Now, I understand well why some activities and programmes that we arrange for patients work and some fail'. It seemed like an 'Aha!' moment for him, processing and recollecting his memories of interacting with patients. As he continued, I pondered the significant role the theory could play in informed decision-making. He described:

> Once I gave a patient a broken clock of mine to fix. I knew back then, before his debilitating mental disease, he was a skilled repair person. The patient eagerly accepted and asked for tools. The next morning, I brought him everything he needed plus a small table with a desk lamp which we placed in the corner of the occupational therapy room and allowed him to attend any time in the day (while we were present) to finish the task. He immersed himself in the activity joyfully. He craved to prove his skills and impress me with the result. He had access to what he needed, plus freedom and control over the timing and way of completing the task (I interpret them as manageability). I remember that he was noticeably more cheerful contrary to many other times he whined about doing compulsory activities that patients should attend weekly. You know, as though he was getting out of bed every morning with a purpose (I call this a meaningful activity specifically for this patient).

In short, the above memory and many other instances showed me how vital it is to educate healthcare practitioners and designers on the theory. This can help purposefully and consciously create a socio-spatial environment supporting occupants' engagement and health. Salutogenesis is not the only theory of health, yet it provides a robust framework with a broad scope encompassing a whole gamut of evidence and

DOI: 10.1201/9781032711195-26

theories on health and design, which allows designers to get a reliable answer – even in complex situations. Salutogenesis can be employed when there is no specific evidence to address the design challenges or when conducting research is hampered by time or budget constraints – the issue that developing countries often face during the projects' design process. Even though salutogenesis has been introduced to various fields (education, health, architectural design of healthcare, aged care, schools, etc.) around the globe, the theory is still under-researched and poorly understood in many developing countries like my country, Iran. I believe one of the underlying challenges hindering the application of theory in architectural design is the insufficient knowledge about salutogenesis among architects and urban designers.

Completing my master's studies and working as an architect, I have always had salutogenic in mind. It has empowered me to ask the right questions throughout the design process and foresee whether a design decision could strengthen the sense of coherence in users or otherwise diminish it. Besides, the salutogenic theory has helped me in my personal life. I am living and working in a country abounding with restrictions and financial and social problems, where one must deal daily with an overwhelming uncertainty about their future. Salutogenesis has helped me stay optimistic in the face of life's challenges. This does not mean that I do not experience moments of despair and disappointment. Indeed, the theory has afforded me the insight to identify where I am standing in the health/illness spectrum, what salutogenic resources are at hand, and what resources I should seek to fortify my senses of comprehensibility and manageability, and enrich meaningfulness to get back on my feet.

23.2 PERSONAL REFLECTIONS BY JAN A. GOLEMBIEWSKI

I have always been sensitive to the physical environment and how it makes me feel. I have a childhood memory of leaving a European marketplace to tour a baroque church nearby. In the market, I wanted stuff: ice cream, sodas, toys. But in the church, I felt numinous. Once the tour ended and we were back in the square, there I was, wanting stuff again – but somehow, I caught myself asking 'why?'. The question has stayed with me all my life. Possibly because I started studying architecture in my early teens, when those ruminations were still so fresh (as a boy I was mentored by Peter Muller OA, an architect and family friend).

Many years later, when I was doing my Masters of Architecture, I started to question the relationship between the emotional states that architecture can so easily trigger and severe mental illness. At first, I wanted to find a soothing language of design that may help people when they were suffering, but as my research flowed into a PhD (now focusing on architecture and schizophrenia), I started finding neuroscientific evidence that the physical environment was more than just good for soothing rattled emotions. There was something in the environment that appeared to be causal (Kapur, 2003), the relevant neural receptors (D2) could be identified, but it was difficult to identify what environmental phenomena activated this transmitter. I published a series of articles while I explored the question – but felt like I was staring down an avalanche of disagreement. At the time, the human genome had just been mapped and mental illness was considered firmly 'medical'. The evidence was being gathered: if any causes were to be found, they were to be genetic or epigenetic. But the smoking gun (gene) was never found. At best, the other scientists found 'susceptibility genes' (e.g., Williams et al., 2007). In 2010, I attended a conference on schizophrenia. The keynote was Professor John McGrath, who was presenting an update on the theory that schizophrenia was caused by vitamin D deficiencies in childhood. Right there and then it occurred to me that it might not be any singular factor in the environment that triggers psychosis, but a lack of experience of the environment as a whole. From the audience, I asked Professor McGrath about whether the cause may be that susceptible people were staying indoors watching TV, and missing critical life experiences, and therefore not getting enough sunlight (vitamin D is synthesised in the skin when it is exposed to UV radiation). His answer was something along the lines of 'You should leave the epidemiology to epidemiologists, and I'll leave architecture for the architects'. Public presentations are stressful places, and I've no doubt he ruminated about better answers afterwards. But I couldn't hear his thoughts. My own theory was presented in the same conference in 2013, and Professor McGrath was among the crowd that was cramming in to hear what I had to say (and to hear his difficult questions!)

It is fine to identify authentic experience as a protective factor against schizophrenia. But the 'why?' question still lingered throughout much of my career. The truth is sometimes difficult to prove, but it is easy to triangulate, so my answers came from diverse modes of inquiry and models. From neuroscience, psychology, and a philosophical model called salutogenics that I stumbled into and discovered a cogent fit. Salutogenics is the theory that the world around us is protective spiritually, emotionally, and physically, and that serious environmental failures mean people become unable to cope and succumb to illness (Golembiewski, 2013). This chapter describes just how and why, and it is illustrated with examples.

REFERENCES

Golembiewski. J. A. (2013, January). Lost in space: The role of the environment in the aetiology of schizophrenia. *Facilities*, *31*(9–10), 427–448.

Kapur, S. (2003). Psychosis as a state of aberrant salience: A framework linking biology, phenomenology, and pharmacology in schizophrenia. *American Journal of Psychiatry*, *160*(1), 13–23. https://doi.org/10.1176/appi.ajp.160.1.13; http://ajp.psychiatryonline.org/cgi/content/abstract/160/1/13

Williams, H. J., Owen, M. J., & O'Donovan, M. C. (2007). Is COMT a susceptibility gene for schizophrenia? *Schizophrenia Bulletin*, *33*(3), 635–641.

Index

Printed in the United States
by Baker & Taylor Publisher Services

Printed in the United States
by Baker & Taylor Publisher Services